# Ambulatory Esophageal pH Monitoring: Practical Approach and Clinical Applications 2nd. Ed.

# Ambulatory Esophageal pH Monitoring: Practical Approach and Clinical Applications 2nd. Ed.

*Joel E. Richter, M.D.*
Chairman, Department of Gastroenterology
The Cleveland Clinic Foundation
Cleveland, Ohio
and
Professor of Internal Medicine
The Cleveland Clinic Foundation
Health Sciences Center of the Ohio State University
Cleveland, Ohio

Williams & Wilkins
A WAVERLY COMPANY

BALTIMORE • PHILADELPHIA • LONDON • PARIS • BANGKOK
HONG KONG • MUNICH • SYDNEY • TOKYO • WROCLAW

*Editor:* Darlene Barela Cooke
*Managing Editor:* Frances M. Klass
*Marketing Manager:* Lorraine A. Smith
*Production Coordinator:* Dana M. Soares
*Typesetter:* Digitype
*Printer:* Quebecor Printing Book Group

*Printed in the United States of America*

Second Edition,

**Library of Congress Cataloging-in-Publication Data**

Ambulatory esophageal pH monitoring : practical approach and clinical
  applications / [edited by] Joel E. Richter. — 2nd ed.
    p.    cm.
  Includes bibliographical references and index.
  ISBN 0-683-30335-X
    1. Gastroesophageal reflux—Diagnosis.   2. Hydrogen-ion
concentration—Measurement.   3. Ambulatory medical care.
  I. Richter, Joel E.
    [DNLM: 1. Esophagus—physiopathology.   2. Gastroesophageal Reflux—
physiopathology.   3. Hydrogen-Ion Concentration.   4. Monitoring,
Physiologic—methods.   5. Ambulatory Care.   WI 250 A497 1997]
  RC815.7.A43   1997
  616.3'3075—DC21
  DNLM/DLC
  for Library of Congress                     97-2003
                                              CIP

ISBN: 0-683-30335-x

*The publishers have made every effort to trace the copyright holders for borrowed material. If they have inadvertently overlooked any, they will be pleased to make the necessary arrangements at the first opportunity.*

97  98  99  00
2  3  4  5  6  7  8  9  10

# Preface for Second Edition

Over the last 25 years, esophageal pH monitoring has grown from a clumsy, confining in-patient study to an ambulatory outpatient procedure performed with compacts solid state technology. This single test is the most sensitive and specific study for identifying gastroesophageal reflux and, more recently, has helped to broaden our horizons about the many extraesophageal presentations of gastroesophageal reflux disease, including chest pain, asthma, hoarseness, coughing, and possibly even laryngeal cancer. This instrumentation has come of age—not only for the gastroenterologist or surgeon, but also for the pulmonary specialist, otolaryngologist, pediatrician, or any physician who sees a large number of patients with difficult to manage gastroesophageal reflux disease.

Although many journal articles have been written about esophageal pH testing, a comprehensive, but practical "how-to-do" it book is not available. Therefore, this book was originally prepared by the editor with the help of many contributing authors with two simple goals in mind: 1) present an understandable practical approach to esophageal pH testing, and 2) review the clinical applications and limitations of this procedure. In developing this project, we have called upon our extensive clinical and research experience with pH monitoring, in some cases spanning nearly 25 years. The section on clinical applications is centered around cases examples and pH tracings to help illustrate the varied uses of esophageal pH monitoring.

The editor and his co-authors are very pleased with the success of the first edition of our book. The second edition expands upon our original theme with the following changes: 1) the addition of foreign contributors from the United Kingdom and The Netherlands, 2) some changes in authorship with six new contributors and 3) expansion on several rapidly evolving areas of gastroesophageal reflux disease including the extraesophageal complications and the controversial area of "alkaline-bile" reflux.

We hope these changes will improve our book so that it continues to be valuable information for everyone who is using esophageal pH testing in either their clinical practice or research environment. It is not a book to be hidden away in the library. If our goal is successful, this book will become worn as it is passed from student to student learning about esophageal pH monitoring.

JOEL E. RICHTER, M.D.

# Acknowledgments

The editor wishes to express his sincere appreciation to the many colleagues who have helped develop our current concepts of esophageal pH monitoring over the years. In addition, I would like to thank Mrs. Rhonda Harris, Ms. Debbie Beam, and Mrs. Susan Fay for their assistance in typing, organizing, and preparing this book.

# Contributors

**Edgar Achkar, M.D.**
Vice-Chairman, Department of
   Gastroenterology
Director, Center for Swallowing
   and Esophageal Disorders
The Cleveland Clinic Foundation
Cleveland, Ohio

**Ross M. Bremner, M.D.**
Resident in Surgery
Department of Surgery
University of Southern California
   School of Medicine
Los Angeles, California

**Donald O. Castell, M.D.**
Professor and Chairman of Medicine
The Graduate Hospital
Philadelphia, Pennsylvania

**Thomas R. DeMeester, M.D.**
Professor and Chairman
Department of Surgery
University of Southern California
   School of Medicine
Los Angeles, California

**Carlo DiLorenzo, M.D.**
Associate Professor of Pediatrics
University of Pittsburgh School
   of Medicine
Pittsburgh, Pennsylvania

**Sarah Douglas**
Technician
Center for Liver and Digestive
   Disorders
The Royal Infirmary of Edinburgh
Edinburgh, Scotland

**Gulchin A. Ergun, M.D.**
Assistant Professor of Medicine
Section of Gastroenterology and
   Hepatology
Northwestern University Medical
   School
Chicago, Illinois

**Susan M. Harding, M.D.**
Assistant Professor of Medicine
Division of Pulmonary and Critical
   Care Medicine
The University of Alabama at
   Birmingham
Birmingham, Alabama

**Robert C. Heading, M.D.**
Center for Liver and Digestive
   Disorders
The Royal Infirmary of Edinburgh
Edinburgh, Scotland

**Paul E. Hyman, M.D.**
Associate Clinical Professor of
 Pediatrics
University of California at
 Los Angeles
Director, Pediatric Gastrointestinal
 Motility Center
Children's Hospital of Orange
 County
Orange, California

**Lawrence F. Johnson, M.D.**
Professor, Department of Medicine
Director, Esophageal Program and
 GI Laboratory
Division of Gastroenterology and
 Hepatology
The University of Alabama at
 Birmingham
Birmingham, Alabama

**Peter J. Kahrilas, M.D.**
Professor of Medicine
Section of Gastroenterology and
 Hepatology
Northwestern University Medical
 School
Chicago, Illinois

**Huitt E. Mattox, III, M.D.**
Gastroenterology Section
Bowman Gray School of Medicine
Wake Forest University
Winston-Salem, North Carolina

**Jean Price, MT (ASCP)**
Laboratory Technician-in-Charge
GI Laboratory
The University of Alabama at
 Birmingham
Birmingham, Alabama

**Cathy A. Schan, PA-C**
Gastroenterology Division
The University of Alabama at
 Birmingham
Birmingham, Alabama

**Andre J.P.M. Smout, M.D.**
Professor of Gastrointestinal
 Motility
Department of Gastroenterology
University Hospital Utrecht
Utrecht, the Netherlands

**Hubert J. Stein, M.D.**
Research Fellow
Department of Surgery
University of Southern California
 School of Medicine
Los Angeles, California

**Kenneth C. Trimble, M.D.**
Consultant Physician and
 Gastroenterologist
Queen Margaret Hospital
Whitefield Road
Dunfermline, Fife Scotland

**Michael F. Vaezi, Ph.D., M.D.**
Department of Gastroenterology
The Cleveland Clinic Foundation
Cleveland, Ohio

**Bas. L.A.M. Weusten, M.D.**
Department of Gastroenterology
University Hospital Utrecht
Utrecht, the Netherlands

# Contents

# 1

# Historical Perspectives on Esophageal pH Monitoring

Lawrence F. Johnson, M.D.

The evolution of 24 hour intraesophageal pH monitoring was the product of a gradual development not only in technology but also in concept. Instrumental to the concept of intraesophageal pH monitoring was the recognition by Winkelstein[1] in 1937 that esophagitis could be caused by reflux of acid—peptic gastric content. This recognition fostered the concept that acid from the stomach could be used as an esophageal marker for reflux. In addition, the commercial availability of in vivo pH probes used in the study of gastric acid secretion further facilitated their use in the study of gastroesophageal reflux, especially since they were equally suitable for placement in the esophagus. The concept that continuous intraesophageal pH monitoring could be used clinically was ultimately ignited by observations such as that acid perfusion induced heartburn in patients with gastroesophageal reflux[2] and that the onset of heartburn coincided with the fall of intraesophageal pH below 4.0.[3] The latter observation was especially pivotal because it equated a common subjective symptom (i.e., heartburn) with a sudden demonstrable change in intraluminal pH.

## EARLY ESOPHAGEAL pH TESTS

Unfortunately, what happened next was a "dipstick" or "litmus paper" approach to pH testing for gastroesophageal reflux. Investigators attempted to "sound" the esophagus with a pH probe, and in a matter of seconds, based upon a specific pH value during insertion or withdrawal, to determine whether the patient was a "refluxer." Known as the Tuttle test,[4] this approach did not succeed because of poor test sensitivity and specificity. Experience with the Tuttle test did, however, provide the impetus to place an indwelling pH probe at a fixed location in the esophagus and instill an acid marker (300 ml of 0.1 nor-

1

mal hydrochloric acid) in the stomach. After instillation of the marker, acid gastro-esophageal reflux was looked for during provocative maneuvers. This test was known as the standard acid reflux test (SART).[5] That the number of reflux events could easily be observed and counted owing to the significant changes in intraluminal pH provided the final impetus to proceed with continuous intraesophageal pH monitoring.

If the standard acid reflux test sparked interest in continuous intraesophageal pH monitoring, why was this technique so long in coming to clinical fruition? Clinical implementation appeared to be held up by two major factors. First, there was a problem with the reference lead. In order to measure intraesophageal pH, a reference lead has to be used in conjunction with the pH probe. Initially, the reference lead consisted of a plastic tube filled with a saturated solution of potassium chloride either as a liquid or as agar. This tube was placed adjacent to the pH probe and had to be maintained in constant contact with the esophageal mucosa in order to attain a successful pH recording. Alternatively, the patient could place a finger in a beaker of saturated solution of potassium chloride adjacent to a glass calomel reference electrode (Beckman No. 402248) or hold this glass electrode in his or her mouth. Obviously, none of these methods worked well or permitted intraesophageal pH monitoring over extended periods of time.

A second factor inhibiting the clinical implementation of continuous intraesophageal pH monitoring was the fact that no investigator had studied a "clean" population of asymptomatic control volunteers. This omission caused the following misperception. When only symptomatic patients with reflux esophagitis were monitored, a commonly held belief arose that acid gastric mucus "stuck to the probe" and caused reflux episodes of long duration. Thus, impaired esophageal acid clearance was perceived as mucus sticking to the pH probe because asymptomatic control volunteers had not been extensively studied. This misperception probably did more to inhibit the development and implementation of continuous intraesophageal pH monitoring than the reference lead problem.

# CONTINUOUS INTRAESOPHAGEAL pH MONITORING

Contrary to popular belief, continuous intraesophageal pH monitoring did not begin during the period from 1969 to 1970 in Great Britain with the studies of Spencer,[6] Pattrick,[7] and Woodward,[8] but instead earlier in the United States by Miller and associates.[9,10] Miller noted that patients with a reflux diathesis of sufficient severity to warrant surgical intervention had reflux episodes of long duration, greater than or equal to 15 minutes. In contrast, asymptomatic controls seldom had reflux events that lasted longer than 5 minutes.[10]

Although they were not the first to perform continuous intraesophageal pH monitoring, Johnson and DeMeester[11] significantly advanced the utilization of this test as an important clinical and investigational tool through their subsequent research and publications. Having just graduated from their respective medical and surgical residencies, both physicians were initially motivated by finding themselves assigned to a military hospital (Tripler Army Medical Center in Honolulu, Hawaii) that cared for an inordinate number of patients with gastroesophageal reflux disease. This assignment occurred at a time (1971 to 1973) when the literature did not advocate any ideal method for evaluating patients with gastroesophageal reflux. After reading about the work from Great Britain,[6-8]

Johnson and DeMeester applied continuous intraesophageal pH monitoring to help solve their patients' clinical problems. After dissatisfaction with the three alternatives for handling the reference lead, the authors found that it would work if placed on the arm. In brief, the point of the reference probe (Beckman No. 402248) was placed in direct contact with the skin, surrounded by the paste used for the electrocardiogram, and wrapped in Saran Wrap to prevent evaporation. This innovation afforded convenience for the patient, almost always ensured a valid pH record, and, most important, permitted extending the pH monitoring period. Since earlier investigators had shown that 12 to 18 hour studies could be tolerated by most patients,[6-8] Johnson and DeMeester decided to monitor the patient for one complete circadian cycle, i.e., 24 hours.[11] This time interval permitted a study that afforded appropriate posturing of the patient during physiologic activities, that is, upright during the day in association with eating three meals and recumbent at night while sleeping.

In order to establish criteria for defining acid reflux abnormalities in symptomatic patients, we studied asymptomatic control volunteers. Our original group consisted of 15 volunteers and made three important contributions. Not only did the controls set standards for judging the degree of abnormality in patients but "physiologic" gastroesophageal reflux was defined for the first time (Figure 1.1). Heretofore, the concept of acid reflux was only considered in a pathophysiologic context. Third, the asymptomatic control population focused attention on the phenomenon of impaired esophageal acid clearance that was occurring in some symptomatic patients but had previously been attributed to acid mucus sticking to the end of the pH probe.

Based on observations from their symptomatic patients, Johnson and DeMeester established six pH criteria upon which to judge the degree of acid gastroesophageal reflux. In

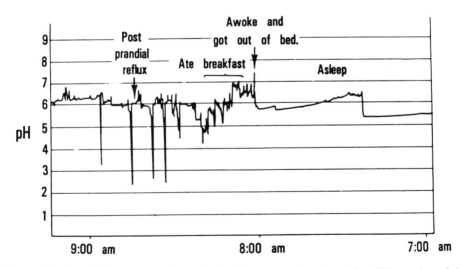

**Figure 1.1.** Physiologic gastroesophageal reflux in an asymptomatic subject. This portion of the pH record shows lower esophageal sphincter competence during sleep and four episodes of reflux after breakfast of which the patient was unaware. Note that this tracing and others to follow reads from right to left as if they were coming directly off a non-ambulatory strip chart recorder in the subjects' room or research laboratory. (From DeMeester TR, Johnson LF, Guy JI, et al: Patterns of gastroesophageal reflux in health and disease. *Ann Surg* 184:459–470, 1976).

addition, they devised a composite score that equated the six different pH criteria into a single value that scored the overall pH record. In brief, the authors provided a "packaged" 24 hour pH monitoring test[11] for the esophageal medical-surgical community (see Chapter 7).

Subsequent studies by Johnson and DeMeester showed that patients with gastroesophageal reflux were not a homogeneous population. When compared with asymptomatic controls, some symptomatic patients had abnormal acid exposure only during the upright period and were called upright refluxers; others had abnormal acid exposure only during the recumbent period and were called recumbent refluxers; and those who were abnormal during both periods were called combined refluxers (Figure 1.2).[12] Reflux during the upright period was characterized by frequent events that were rapidly cleared (Figure 1.3) and reflux during the recumbent period was characterized by infrequent events that were slowly cleared (Figure 1.4). With different reflux patterns, these three patient groups provided the opportunity to study which pattern(s) caused reflux esophagitis. It was found that patient groups with abnormal recumbent distal esophageal acid exposure seemed predisposed to the highest incidence and most severe forms of esophagitis noted at endoscopic examination (Figure 1.5).[12] In contrast, upright refluxers were found to have a low incidence of reflux esophagitis, despite the frequency of their reflux events. To test the validity of this observation by other means, we compared the effect of the reflux patterns from the same three patient groups as well as from patients with normal 24

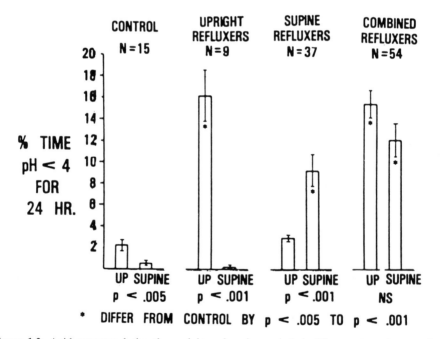

**Figure 1.2.** Acid exposure during the upright and supine periods in 15 asymptomatic normal volunteers and 100 patients with symptoms of gastroesophageal reflux. (From DeMeester TR, Johnson LF, Guy JI, et al.: Patterns of gastroesophageal reflux in health and disease. *Ann Surg* 184:459–470, 1976.)

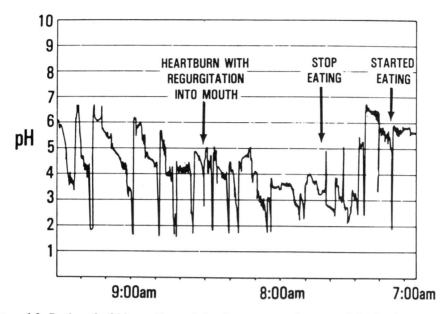

**Figure 1.3.** Portion of a 24 hour pH record showing exaggerated postprandial reflux in a symptomatic patient during the upright period. (From DeMeester TR, Johnson, LF, Guy JI, et al.: Patterns of gastroesophageal reflux in health and disease. *Ann Surg* 184:459–470, 1976.)

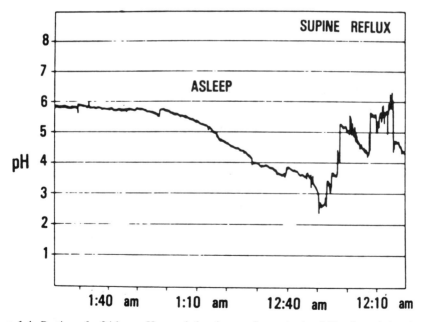

**Figure 1.4.** Portion of a 24 hour pH record showing a reflux episode of 30 minutes' duration that occurred while the patient was asleep. (From DeMeester TR, Johnson, LF, Guy JI, et al: Patterns of gastroesophageal reflux in health and disease. *Ann Surg* 184:459–470, 1976.)

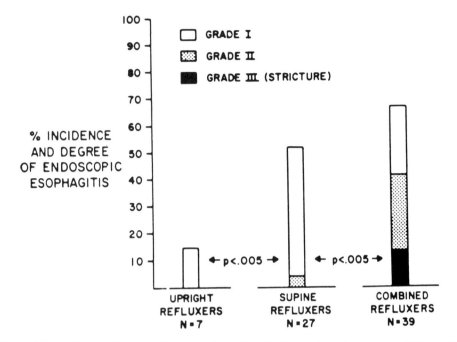

**Figure 1.5.** Incidence and degree of endoscopic esophagitis in symptomatic patients with three different patterns of excessive acid exposure. (From DeMeester TR, Johnson, LF, Guy JI, et al: Patterns of gastroesophageal reflux in health and disease. *Ann Surg* 184:459–470, 1976.)

hour pH scores as to the degree of reactive epithelial change consistent with reflux esophagitis noted on their biopsies (Figure 1.6).[13] Reflux esophagitis (papillary length greater than 60 percent) appeared again to occur in those groups with abnormal recumbent acid exposure. This exposure occurred as a result of more frequent reflux events and a posture-related impairment in the esophageal acid clearance time.[13] The frequency of upright reflux events (number per hour), the duration of upright acid exposure (minutes), and even the duration of the total period of acid exposure (minutes) were not found to be strong determinants of reflux esophagitis (see Figure 1.6).[13] Thus, the risk factors for the reflux esophagitis seemed to be the timing and duration of reflux episodes. Clearly, the nocturnal period appeared to be the most detrimental time for reflux to occur, especially if there was a posture-related impairment in the nocturnal esophageal acid clearance time.[12,13] Interestingly, hiatus hernias were also shown to delay the nocturnal esophageal acid clearance time.[13] In addition to affecting esophagitis, the reflux patterns in the three patient groups had clinical relevance, especially since upright refluxers were shown to have increased problems with gas bloat syndrome after fundoplication, but could respond to unorthodox therapy such as biofeedback.[15] Among their other early work, Johnson and DeMeester also showed that 24 hour intraesophageal pH monitoring could be used to measure objectively responses to either medical or surgical therapies.[14,16,17]

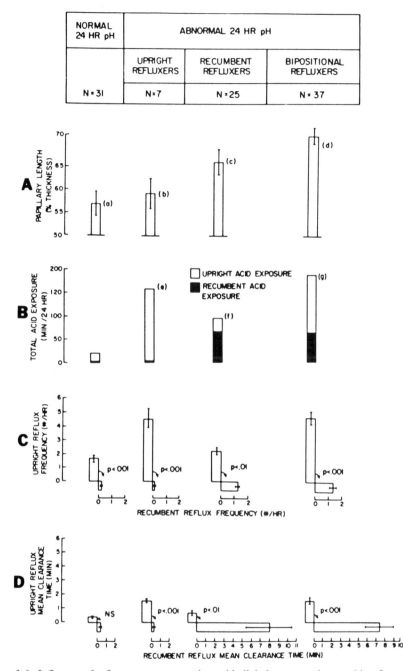

| NORMAL 24 HR pH | ABNORMAL 24 HR pH | | |
|---|---|---|---|
| | UPRIGHT REFLUXERS | RECUMBENT REFLUXERS | BIPOSITIONAL REFLUXERS |
| N=31 | N=7 | N=25 | N=37 |

**Figure 1.6.** Influence of reflux pattern on reactive epithelial change consistent with reflux esophagitis (percent papillary extension). $N$ = number of patients. All bars show mean and $\pm$ 1 standard error of the mean. Minutes of total acid exposure shown by the entire bar; unshaded area equals minutes upright; shaded area equals minutes recumbent. Upright reflux frequency and acid clearance time are shown on vertical bars and comparable recumbent parameters on horizontal bars. The sign ($\longrightarrow$) and adjacent "$p$" denote significance of change for that parameter after change from the upright to recumbent posture in each patient group. Papillary length (d) and (c) both more than (a), with $p < 0.001$ and 0.05, respectively; (d) more than (b), with borderline significance ($0.1 > p > 0.05$). Total acid exposure (f) is less than both (g) ($p < 0.001$) and (e) ($0.1 > p > 0.05$). NS = not statistically significant. (From Johnson LF, DeMeester TR, Haggitt RC. Esophageal epithelial response to gastroesophageal reflux: A quantitative study. *Am J Dig Dis* 23:498–509, 1978.)

# AMBULATORY OUTPATIENT INTRAESOPHAGEAL pH MONITORING

In the early 1980s, continuous intraesophageal pH monitoring underwent another dramatic change. This time the change was motivated by our health care system as it was transformed from in-hospital based diagnostic testing to ambulatory outpatient diagnostic testing. A second motivating factor was miniaturization of electrical equipment such as pH meters, recorders, and computers. Thus, ambulatory outpatient intraesophageal pH monitoring was born by affording more comfortable and compact systems that permitted patients to be studied in their own homes or work environments.

Although ambulatory outpatient intraesophageal pH monitoring obviated the hospital expense and freed the patient from a sedentary controlled environment, it introduced both the patient and investigating physician to an uncontrolled ambulatory outpatient environment. The variables in this new environment included diversity in the test diet, variations in meal timing, differing physical activity levels, and poor adherence to instructions regarding posture and sleep during the day. These variables, along with the difficulty of maintaining quality control of the monitoring equipment outside the hospital, were probably responsible for increases in the values for the different pH criteria noted in asymptomatic control volunteers.[18]

Despite the increase in control values,[18] it is surprising how well values determined by in-hospital pH monitoring during the period 1972–1974[11] compared with those attained either in the hospital[19] or in an ambulatory state out of the hospital[20–22] by authors from the United States,[20,21] Spain,[19] and Italy[22] during the period from 1988 to 1990 (Table 1.1). In addition, it is prudent to note that the concept of physiologic gastroesophageal reflux remained constant. That is, acid exposure (percentage of time pH was less than 4) noted while the patient was awake in the upright posture always exceed that while the patient was asleep in the recumbent posture.

Several studies[23,24] have shown that the high tech ambulatory systems that take readings at various time intervals (usually every 6 to 8 seconds) accurately measure acid reflux and clearance compared with real time measurements obtained by the old pH meters and strip chart recorders. Thus, it is apparent in the history of continuous intraesophageal pH monitoring that an important milestone has been attained. This attainment represents a successful conversion from a sedentary, controlled environment of in-hospital pH monitoring that uses standard equipment to that of a less than ideally controlled ambulatory outpatient environment that uses equipment specifically engineered for this purpose.

Today, one should choose ambulatory outpatient intraesophageal pH monitoring to evaluate a variety of symptoms in the setting of the patient's home or work environment. This setting is ideal, especially since gastroesophageal reflux tends to increase in the ambulatory environment as opposed to the controlled sedentary environment of the hospital. However, the reader should be aware that the ambulatory environment and the new equipment do not appear to improve the diagnostic accuracy of this test over that noted for in-hospital pH monitoring.[22] This lack of improvement probably results because the poorly controlled nature of the ambulatory environment increases gastroesophageal reflux in the asymptomatic control volunteers against whom the patients are judged.

**TABLE 1.1. ASYMPTOMATIC CONTROL VALUES FOR 24 HOUR ESOPHAGEAL
pH MONITORING**

| pH Criteria Mean ± 1 SD | Authors | | | | |
|---|---|---|---|---|---|
| | Johnson N = 15 1974[11] | Zaninotto N = 50 1988[21] | Pujol N = 15 1988[19] | Mattioli N = 20 1989[22] | Mattox N = 48 1990[20] |
| Upright (%) | 2.3 ± 2.0 | 2.4 ± 2.4 | 2.2 ± 1.4 | 2.8 ± 2.4 | 2.2 ± 2.3 |
| Supine (%) | .29 ± .46 | .55 ± 1.0 | .20 ± .20 | .66 ± .80 | .50 ± .77 |
| Total (%) | 1.5 ± 1.4 | 1.6 ± 1.5 | 1.3 ± .70 | 1.9 ± 1.6 | 1.6 ± 1.5 |
| No. of reflux events | 20.6 ± 14 | 23.8 ± 23 | 20.0 ± 14 | * | 20.1 ± 15 |
| No. of reflux events ≥5 | 0.6 ± 1.2 | .95 ± 1.3 | 0.3 ± 0.6 | * | 0.6 ± 1.3 |
| Longest reflux event | 3.9 ± 2.7 | 6.8 ± 8.2 | 3.5 ± 2.7 | 5.2 ± 5.0 | 4.5 ± 5.6 |

*Denotes mean values given as reflux events per hour. When multiplied by 24 hours, number of reflux events and number of reflux events ≥5 minutes were 24 and 0.7, respectively.

# RESEARCH STIMULATED BY 24 HOUR INTRAESOPHAGEAL pH MONITORING

A good research technique sparks additional investigation that uses a different approach for resolving its unanswered questions, and 24-hour intraesophageal pH monitoring proved no exception. Observations attained from the 24 hour pH records of patients have been applied to an in vivo animal model (rabbit esophagus). These studies were conducted over almost 10 years[25] and along with those of others helped to form the hypothesis that the proteolytic properties of pepsin constitute a severe injurious agent in reflux esophagitis. Moreover, histamine 2 blockers and now the proton pump inhibitors appear to exert their healing effect on esophagitis by raising the intraluminal pH above the optimal pH for pepsin (greater than or equal to 4.5), thereby destroying its injurious properties. This hypothesis is supported by the observation that acid alone at pH values observed in humans during 24 hour pH monitoring does not appear to be that injurious in the animal model.

Another line of investigation sparked by 24 hour intraesophageal pH monitoring was the interrelationship between the physiologic process of sleep and esophageal acid clearance. These studies were conducted in the control environment of a sleep laboratory with simultaneous electroencephalographic and intraesophageal pH monitoring.[26,27] In brief, the studies showed that sleep prolonged the esophageal acid clearance time, a nonproblem in asymptomatic controls who rarely have reflux at night but the first of a double jeopardy phenomenon in symptomatic patients with reflux esophagitis. The second jeopardy consisted of a delay in the esophageal acid clearance time that occurred despite comparable arousals from sleep and swallowing frequency when compared with controls. This delay suggested a primary peristaltic defect in patients with reflux esophagitis that has now been documented by multiple groups. Thus, these sleep laboratory and animal studies are but a few of the many instances in which observations from 24 hour intraesophageal pH monitoring yielded insightful and clinically relevant results when followed up by other research techniques.

## THE FUTURE

The future for continuous intraesophageal pH monitoring appears bright, especially since it is now recognized as an established technique to (1) study esophageal motility with respect to acid reflux, (2) document decreases in acid exposure fostered by therapeutic interventions, (3) determine the proper dose and administration schedules for new drugs that treat reflux esophagitis, and (4) preselect specific patient populations for clinical study. Moreover, this technique can also be used for studying the relationship between gastroesophageal reflux and disorders such as bronchial asthma, ear, nose and throat disorders or chest pain. Recently, multisite pH monitoring (i.e., proximal and distal esophagus) has helped elucidate the risk factors connected with reflux-associated respiratory and laryngeal disease. Finally, simultaneous esophageal and gastric pH monitoring has documented the phenomenon of duodenogastric reflux and how it affects the pH of the esophageal refluxate. Hence, both now and in the future, there are many old and new applications for continuous intraesophageal pH monitoring.

## REFERENCES

1. Winkelstein A: Peptic esophagitis: A new clinical entity. *JAMA* 104:906–912, 1935.
2. Bernstein LM, Baker LA: A clinical test for esophagitis. *Gastroenterology* 34:760–781, 1968.
3. Tuttle SG, Rufin F, Battaneloo A: The physiology of heartburn. *Ann Intern Med* 55:292–300, 1961.
4. Tuttle SG, Grossman MI: Detection of gastroesophageal reflux by simultaneous measurements of intraluminal pressure and pH. *Proc Soc Exp Biol Med* 98:224, 1958.
5. Kantrowitz PA, Carson JG, Fleischli DG, et al: Measurement of gastroesophageal reflux. *Gastroenterology* 56:666–674, 1969.
6. Spencer J: Prolonged pH recording in the study of gastro-esophageal reflux. *Br J Surg* 56: 912–914, 1969.
7. Pattrick FG: Investigation of gastroesophageal reflux in various positions with a two-lumen pH electrode. *Gut* 11:659–667, 1970.
8. Woodward DAK: Response of the gullet to gastric reflux in patients with hiatal hernia and oesophagitis. *Thorax* 24:459–464, 1970.
9. Miller FA: Utilization of inlying pH probe for evaluation of acid peptic diathesis. *Arch Surg* 89:199–203, 1964.
10. Miller FA, Doberneck RC: Diagnosis of the acid-peptic diathesis by continuous pH analysis. *Surg Clin North Am* 47:1325–34, 1967.
11. Johnson LF, DeMeester TR: Twenty-four hour pH monitoring of the distal esophagus: A quantitative measure of gastro-esophageal reflux. *Am J Gastroenterol* 62:325–332, 1974.
12. DeMeester TR, Johnson LF, Guy JI, et al: Patterns of gastroesophageal reflux in health and disease. *Ann Surg* 184:459–470, 1976.
13. Johnson LF, DeMeester TR, Haggitt RC: Esophageal epithelial response to gastroesophageal reflux: A quantitative study. *Am J Dig Dis* 23:498–509, 1978.
14. DeMeester TR, Johnson LF, Kent AH: Evaluation of current operations for the prevention of gastroesophageal reflux. *Ann Surg* 180:511–525, 1974.
15. Shay SS, Johnson LF, Wong RKH, et al: Rumination, heartburn and daytime gastroesophageal reflux: A case study with mechanisms defined and successfully treated with biofeedback therapy. *J Clin Gastroenterol* 8:115–126, 1986.

16. DeMeester TR, Johnson LF: Evaluation of the Nissen antireflux procedure by esophageal manometry and twenty-four hour pH monitoring. *Am J Surg* 129:94–100, 1975.

17. Johnson LF, DeMeester TR: Evaluation of elevation of the head of the bed, bethanechol, and antacid foam tablets on gastroesophageal reflux. *Dig Dis Sci* 26:673–680, 1981.

18. Branicki FJ, Evans DF, Ogilvie AL, et al: Ambulatory monitoring of oesophageal pH in reflux oesophagitis using a portable radiotelemetry system. *Gut* 23:992–998, 1982.

19. Pujol A, Grande L, Ros E, et al: Utility of inpatient 24-hour intraesophageal pH monitoring in diagnosis of gastroesophageal reflux. *Dig Dis Sci* 33:1134–1140, 1988.

20. Mattox HE, Richter JE. Prolonged ambulatory esophageal pH monitoring in the evaluation of gastroesophageal reflux disease. *Am J Med* 89:345–356, 1990.

21. Zaninotto G, DeMeester TR, Schwizer W, et al: The lower esophageal sphincter in health and disease. *Am J Surg* 155:194–201, 1988.

22. Mattioli S, Pilotti V, Spangaro M, et al: Reliability of 24-hour home esophageal pH monitoring in diagnosis of gastroesophageal reflux. *Dig Dis Sci* 34:71–78, 1989.

23. Herrera JL, Simpson JK, Wong RK, et al: Comparison of stationary vs ambulatory 24-hour pH monitoring recording systems. *Dig Dis Sci* 33:385–388, 1988.

24. Allen ML, Orr WC, Woodruff DM, et al: Validation of an ambulatory esophageal pH monitoring system. *Am J Gastroenterol* 83:287–290, 1988.

25. Johnson LF, Harmon JW: Experimental esophagitis in a rabbit model: Clinical relevance. *J Clin Gastroenterol* 8:26–44, 1986.

26. Orr WC, Robinson MG, Johnson LF: Acid clearing during sleep in patients with esophagitis and controls. *Dig Dis Sci* 26:423–427, 1981.

27. Orr WC, Johnson LF, Robinson MG. The effect of sleep on swallowing, esophageal peristalsis, acid sensitivity and clearance time. *Gastroenterology* 86:814–819, 1984.

# 2

# Ambulatory Esophageal pH Monitoring: Electrodes and Recording Equipment

**Kenneth C. Trimble, M.D.**
**Sara Douglas, Bsc.**
**Robert C. Heading, M.D.**

Ambulatory pH monitoring affords an opportunity to detect the occurrence, frequency, and duration of the pathophysiological events that characterize the disease process. It investigates, and allows the coincidence of the patients symptoms with the events to be quantified instead of just demonstrating the damaging effects on the end-organ. In this important respect it differs from other modalities commonly used to investigate gastroesophageal reflux disease (GERD), such as radiologic or endoscopic methods.[1] In addition, the development of the technique has dramatically enhanced our understanding of GERD, not least in an appreciation of the wide spectrum of the disease, and that many patients suffering symptoms from GERD do not have detectable end-organ damage.[2] The need for an objective means of measuring reflux and assessing its relation to symptoms has long been recognized, but not until the late 1970s and early 1980s, with the introduction of truly ambulatory prolonged esophageal pH monitoring,[3] did an acceptably sensitive and specific tool become available. Earlier studies using nonportable equipment were limited to the in-patient setting. They compromised the accessibility and relevance to real life situations in which most patients experience reflux symptoms. The technological advances over the last two decades have greatly simplified the procedure, not only with the development of lightweight portable recorders allowing out-patient ambulatory studies, but also in the processing and presentation of the huge amounts of information generated by 24 hour studies. Consequently, the technique has progressed from an exclusively laboratory-based research tool to a routine out-patient "office" clinical investigation. We discuss the current practice of ambulatory pH monitoring in terms of the equipment currently available and its use. It is not our intention to recommend or dissuade, but merely to highlight the important points to be considered when setting up and carrying out ambulatory pH monitoring.

# PRINCIPLES OF pH MEASUREMENT

Like many clinical diagnostic procedures, ambulatory pH monitoring has been subject to great technological advances[4,5] to the point where a large range of compact, accurate equipment is now readily available. Notwithstanding, these progressions have remained true to the electrochemical basis of pH measurement i.e., production of a galvanic cell.[6] In any galvanic cell, there are two electrodes, one of which is a reference electrode with a constant potential. The other is called an indicating electrode: the ion selective electrode whose potential is sensitive to the concentration of the species of interest. Both electrodes are connected to a device that translates potential into concentration information. Thus the pH of a solution can be determined, where one pH unit is equivalent to 62 mV of potential difference.[7] The minimum equipment requirement for ambulatory pH monitoring is connection of the electrode to a suitable recorder with the ability to translate and store 24 hours worth of data.

# pH ELECTRODES

pH sensors or 'electrodes' exist in several forms, of which the two most common are glass and antimony (Figure 2.1). The further differentiation into monopolar electrodes

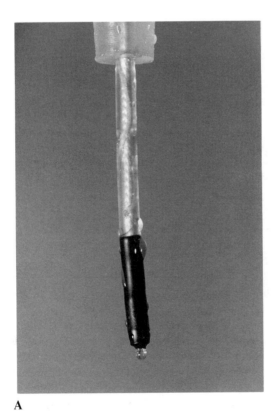

A

**Figure 2.1.** (a) Combined glass electrode.

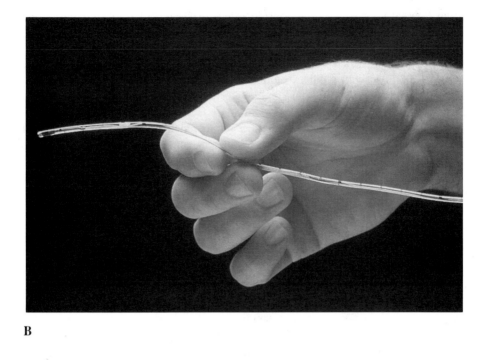

B

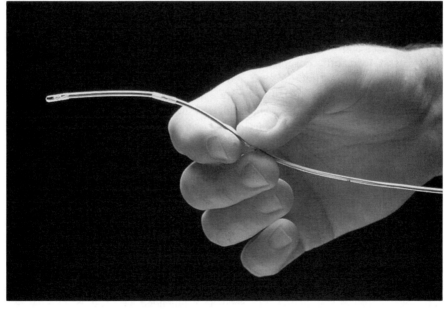

C

**Figure 2.1.** (b) Semi-disposable monopolar antimony electrode with single pH sensor. (c) Semi-disposable monopolar antimony electrode with dual pH sensors.

with an external reference and combination electrodes with a built-in reference is also possible. There are important differences between these two types.

Glass electrodes are generally considered to be the most accurate of the sensors available.[8] They are constructed from a hollow glass tube sealed with an ion sensitive glass membrane and filled with a dilute solution of HCl. Into this solution is dipped a silver wire coated with silver chloride. In combination with a reference (calomel) electrode, a cell is created and the resulting potential difference is dependent only on the external solution. The reference electrode may be either intraluminal (preferred) or a less accurate cutaneous electrode. The susceptibility of the reference electrode to dislodgment in addition to unavoidable fluctuations in skin pH make it a common source of artifact that can render a study uninterpretable. Glass electrodes in the form of radiotelemetry capsules, which are swallowed and tethered to the cheek with nylon thread, are also available. This method is not widely used, although recordings comparable with that of more conventional electrodes have been documented.[4]

Clinical studies require a sensor that is both economical and reliable. Glass electrodes are relatively expensive and need careful handling. Strong sterilizing solutions such as gluteraldehyde, although necessary, impair their function and exposure times in excess of 20 minutes are not recommended. A thorough rinsing with sterile water for **at least 30 minutes** is necessary after glutaraldehyde sterilization. Prolonged exposure to acidic solutions also accelerate degeneration of the internal cell and must only be used for calibration purposes. The electrodes must be stored upright, in protective tubes containing an electrolyte solution (normally supplied by the manufacturer). This will allow the maximum number of studies to take place and safeguard against failure *in situ*. In our laboratory, we can expect a range of 55–65 uses for a glass electrode (approximately a 6 month lifespan). Electrode failure is easily determined from the resulting pH tracing. Wild pH fluctuations from acid to alkaline or off-scale values are common indications, as is unresponsiveness to foodstuffs consumed by the patient.

Although considerably more robust and a good deal cheaper, monocrystalline antimony electrodes have a much shorter lifespan and may only be used for an average of 5–10 twenty-four hour recordings. At present they are only available with an external reference and although this is seen as a disadvantage, the resulting probe is smaller, facilitating trans-nasal insertion, especially useful in infants and children.[9] pH is measured by corrosion of the crystal within the body of the electrode, thus resulting in a finite lifespan, which in economical terms, decreases their usefulness. They can be used satisfactorily in the clinical setting however, but only for esophageal recordings.

Unlike glass, antimony electrodes are not restricted to one pH sensor per catheter (Figure 2.1c). Dual channel pH catheters allowing gastro-esophageal or pharyngo-esophageal studies are now available, and although this use has not yet been validated, recent studies have investigated connections between cervical symptoms and proximal acidification.[10] Elevated acid levels have also been recorded proximally in patients presenting with Barrett's esophagus.[11] Gastric pH monitoring may be useful in the interpretation of negative esophageal recordings,[12] although this is not recommended as a routine clinical procedure.

Studies have shown that in terms of sensitivity, response time, pH drift, and temperature coefficient antimony electrodes are clearly inferior to glass[13] (Figure 2.2). In addition their response curves show considerable hysteresis. From a research standpoint they are unsuitable for any accurate pH measurement.[13,14]

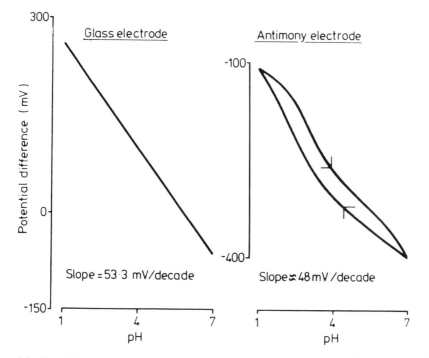

**Figure 2.2.** Sensitivity of a representative glass and an antimony electrode. Note that the glass electrode has a linear response, which is not the case with the antimony electrode. In fact, a different response curve is found when moving down compared with moving up a pH gradient, i.e., hysteresis. (From McLaughlan, G, Rawlings JM, Lucas MT, et al: Electrodes for 24 hour pH monitoring—a comparative study. *Gut* 28:1935, 1987, with permission.)

A recent introduction to the sensors available has been the ion-sensitive field effect (ISFET) pH electrode. This is a modification of a field effect transistor used in electronic circuitry and utilizes an ion-sensitive membrane to produce a voltage proportional to pH. Currently available only as monopolar, they appear to be comparable to glass electrodes with respect to linearity and drift but respond faster to sudden pH changes. They can be miniaturized and several sensors can be incorporated into one catheter and may represent a major advance in intraluminal pH measurement.[15,16]

# ELECTRODE CALIBRATION

Prior to each study, an in vitro two point calibration (usually pH 1 and 7) must be carried out. The electrode and reference (where applicable), are placed in each buffer solution at either room or body temperature (37°C) until stabilization is reached. This calibration should be repeated on return of the patient to rule out electrode failure and to check for slow pH drift. A drift of less than 0.5 pH units over the entire 24 hour period is considered acceptable. With respect to temperature, reference electrodes, and connection jelly, mathematical correction factors must be included in the pH calculation as these can vary the final measurement greatly. Many of the software packages available today however, combine these into one multifactorial correction value for automatic calculation. This is an extremely useful and time-saving feature that enhances the desirability of computer-aided pH measurement.

# ELECTRODE POSITIONING

By convention the pH electrode (or probe) is positioned 5 cm above the proximal border of the lower esophageal sphincter (LES). This avoids displacement of the electrode into the stomach during the shortening of the esophagus that occurs on swallowing. Determination of the position of the LES by means of a standard stationary esophageal manometry study is generally regarded as the optimum method for pH probe localization[5,8] (See Chapter 4 for further discussion). Further advances in technology have allowed for a much greater memory capacity without increase in size or weight of the recorder, so that multichannel pH tracings or combined pH and pressure recordings can be produced through addition of a pressure sensor to the pH probe. This does increase costs and probe diameter, but the benefits of this method has led to its increasing popularity where only a pH study is required[17] (Figure 2.3).

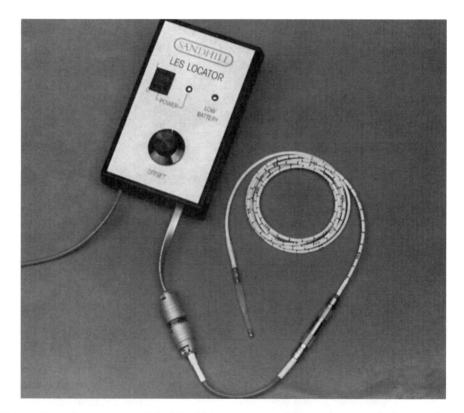

**Figure 2.3.** LES Locator (Sandhill Scientific). A pressure microtransducer attached to a probe. A pull-through can be done with these assemblies to locate the lower esophageal sphincter and to place the pH probe 5 cm above the proximal margin of the lower esophageal sphincter. This technology eliminates the need for two intubations to properly place the pH probe.

A multi-sensor pressure catheter and combined pH electrode resulting in whole esophagus, simultaneous 24 hour recordings via a slimline, well-tolerated catheter are a logical progression and would allow greater insight into the more subtle motility disorders for which conventional manometric investigation may be inadequate for the identification of abnormality.

# RECORDING AND ANALYSIS OF DATA

Twenty-four hour pH monitoring produces a wealth of information about changes in esophageal pH, and as a result of the development of computer analysis, variables regarding number, frequency, and duration of reflux episodes in addition to derived calculations can be readily generated. The newer software packages have become progressively easier to use, and allow for a very flexible display format for the pH tracing in compressed or expanded forms, as well as annotation of patient events from the symptom diary or event marker (Figure 2.4). In addition to the directly measured variables such as total acid exposure time and episode duration, available software now allows the automatic generation of calculated variables such as composite severity scores (e.g., the DeMeester score[18]) and measures of symptom correlation such as the symptom index.[19] Both the display and expression of measured and derived parameters can be customized to the individual users requirements, and data summaries and reports readily produced (Figure 2.5). Some of the software packages also allow for considerable mathematical transformation of the data so that, for example, pH distribution curves can be plotted. Not only has computer software rendered the analysis of pH data much quicker and simpler, but the database facilities greatly enhance archiving and audit.

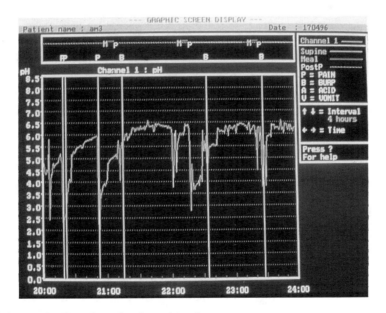

**Figure 2.4.** Sample pH tracing using Synectics software.

NUMERICAL ANALYSIS REPORT - an3

| ACID REFLUX | | Total | Uprght | Supine | Meal | PostP |
|---|---|---|---|---|---|---|
| Duration | (HH:MM) | 13:30 | 06:51 | 06:39 | 01:10 | 07:00 |
| Number of reflux episodes | ( # ) | 26 | 24 | 2 | 7 | 18 |
| Number of reflux episodes longer than 5.0 minutes | ( # ) | 3 | 2 | 1 | 0 | 2 |
| Longest reflux episode | (min) | 13 | 6 | 13 | 1 | 6 |
| Total time pH below 4.00 | (min) | 47 | 34 | 13 | 3 | 31 |
| Fraction time pH below 4.00 | ( % ) | 5.8 | 8.2 | 3.3 | 4.0 | 7.4 |

| ACID REFLUX | | PAIN | BURP |
|---|---|---|---|
| Duration | (HH:MM) | 00:20 | 00:44 |
| Number of reflux episodes | ( # ) | 6 | 4 |
| Number of reflux episodes longer than 5.0 minutes | ( # ) | 0 | 0 |
| Longest reflux episode | (min) | 4 | 2 |
| Total time pH below 4.00 | (min) | 8 | 3 |
| Fraction time pH below 4.00 | ( % ) | 40.7 | 7.4 |

Use ↑↓ <PgDn> and <PgUp> to move. <Esc> to leave

**Figure 2.5.** Sample pH tabular results.

# DATA RECORDERS

In the early days of ambulatory recordings, portable magnetic cassette tape recorders, were used that could be replayed onto a chart recorder and the data analyzed manually. The development of small, lightweight solid state digital recorders (Figure 2.6) has resulted in the ambulatory pH monitoring and analysis becoming much more accessible and easier for the patient (Figure 2.7).

Unlike direct pH recording that is continuous, compact data loggers only sample pH intermittently. A rate of eight samples/minute or more is sufficiently accurate for measurement of median and threshold pH values for clinical purposes.[20] All the commercially available compact recorders satisfy this criterion (Figure 2.8). Sampling rates can also be varied on many of the recorders. Further advances in technology have allowed for a much greater memory capacity without increase in size or weight of the recorder, so that multichannel pH tracings or combined pH and pressure recordings can be produced. Currently available data loggers allow for up to 96 hours of recording sampled once every 6 seconds from recorders weighing as little as 300g.

All recorders feature various patient event buttons to record symptoms or events during the recording and allow subsequent correlation with reflux events. Some of these can be preprogrammed to label particular symptoms or activities (e.g., chest pain, heartburn, smoking). It is advisable, however, for patients to keep their own simple written record, as the use of event markers, especially if multiple, can be confusing for some patients.

Most of the recorders currently on the market fulfill similar criteria and differ in mem-

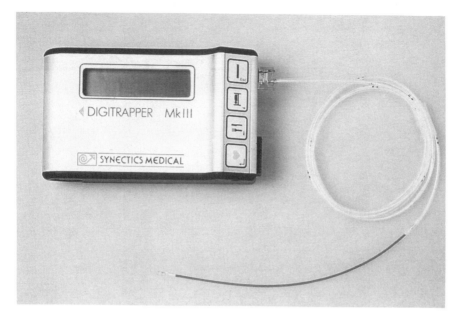

A

B

**Figure 2.6.** Portable digital recorders: a) Synectics Medical, Irving, Texas USA, b) Flexilog 3000, Oakfield Instruments Ltd., Witney, Oxon UK.

**Figure 2.7.** Patient with recording equipment in place.

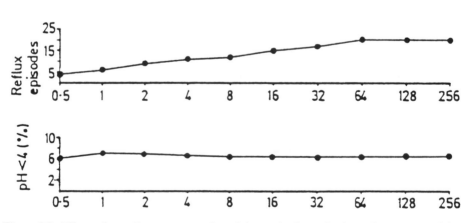

**Figure 2.8.** Effects of sampling rate on number of detected reflux episodes and percentage of time pH is less than 4.0. (Modified from Emde C, Garner A, Blum AL. Technical aspects of intraluminal pH-metry in man: Currrent status and recommendations. *Gut* 28:1177, 1987, with permission.)

ory capacity, data storage methods, and system configuration flexibility. Those recorders that incorporate removable memory cards, allow instant memory extension and increased data storage capacity. This method however, may be more expensive and recorders with a built-in set memory are perfectly adequate for most studies. Many manufacturers are now adopting a 'minimalist' approach to their designs, creating compact, portable recorders with the minimum number of function buttons, requiring relatively few specialist skills. Recorders should be compatible with both glass and antimony electrodes as standard and come supplied with a suitable carrying pouch.

## RECORDING SOFTWARE

Since the introduction of digital recorders, computers and the associated software required, have become an almost indispensable tool in ambulatory pH monitoring. The versatility and time-saving features of the systems available allow recordings of greater accuracy, produced in a much shorter timespan. High memory, multichannel recorders, used in conjunction with user-friendly, compatible software has sent ambulatory pH monitoring into a new dimension. Packages currently on sale now allow the user to produce the standard range of analyses designed for clinical diagnosis plus composite scoring values e.g., the Demeester's.[18] Statistical calculations such as frequency distribution (Figure 2.9) and circadian rhythm diagrams are now included in a number of available systems, and use of these is at the investigators' discretion.

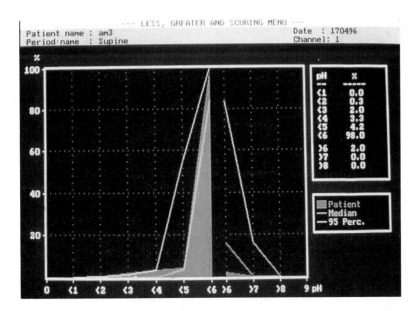

**Figure 2.9.** Sample alternative data displays.

Developments in the computer business have also heralded changes for software users. Windows™ based packages incorporating mouse controlled functions and color graphics that produce an aesthetic, comprehensible presentation of results are generally regarded as the way forward by many manufacturers. The standard installation of Windows-type software into new personal computers has allowed investigators to become comfortable with its use, thus enabling custom designed reports to be produced simply and efficiently.

Each manufacturer provides the software for use with their system and these programs are generally incompatible with any other. Different centers use different values for their calculations and whether a Windows or DOS based system is used, the software must be flexible in its set up and allow easy data manipulation. Choice of software should take into account the computer system available, (many manufacturers will supply a personal computer along with the recording equipment for an extra charge) and the needs of the user. Software upgrades occur frequently but are not necessarily an advantage.

# CONCLUSION

Improved technological development has greatly increased the ease with which ambulatory esophageal pH recording can be performed on out-patients with suspected GERD, and rendered the interpretation of the data so obtained much less laborious and time-consuming. As a result, this most sensitive and specific test for the diagnosis of one of the commonest gastrointestinal disorders should come within the reach of most practicing GI clinicians.

# REFERENCES

1. Smout A: Ambulatory monitoring of esophageal pH and pressure. In Castell DO, ed: *The Esophagus.* 2nd ed. Little, Brown, 1995, 153.
2. Johnsson F, Joelsson B, Gudmundsson K, Grieff L: Symptoms and findings in the diagnosis of gastro-oesophageal reflux disease. *Scand J Gastroenterol* 22:714, 1987.
3. Johnson LF, De Meester TR: Twenty-four hour pH monitoring of the distal esophagus. A quantitative measure of gastresophageal reflux. *Am J Gastroenterol* 62:325, 1974.
4. Branicki FJ, Evans DJ, Ogilvie AL: Ambulatory monitoring of oesophageal pH in reflux oesophagitis using a portable radiotlemetry system. Gut 23:992, 1982.
5. Bennett JR: pH measurement in the oesophagus. In Tytgat GNJ, ed: *Oesophageal Disorders.* London, Bailliere Tindall, 1987, 747.
6. De Meester TR, Johnson LF, Joseph GJ, Toscano MS, Hall AW, Skinner DB: Patterns of gastroesophageal reflux in health and disease. *Ann Surg* 184:459, 1976.
7. Rovelstad RH, Owen CA, Magrath TB: Factors influencing the continuous recording of in situ pH of gastroduodenal contents. *Gastroenterology* 83:629, 1988.
8. De Caestecker JS, Heading RC: Esophageal pH monitoring. *Gastroenterol Clin North Am* 19:645, 1990.
9. McLaughlin G, Rawlings JM, Lucas ML: Electrodes for 24 hour pH-metry: a comparative study. *Gut* 28:935, 1987.
10. Ward BW: Ambulatory 24 hour oesophageal pH monitoring: technology searching for a clinical application. *J Clin Gastroenterol* 8:59, 1986.

11. Duroux P: The ion sensitive field effect transistor (ISFET) pH electrode: a new sensor for long term ambulatory pH monitoring. *Gut* 32:240, 1991.

12. Kuit JA: Evaluation of a new catheter for esophageal pH monitoring. *Hepatogastroenterology,* 38:78, 1991.

13. Galmiche JP, Scarpignato C: Esophageal pH monitoring. In Scarpignato C, Galmiche JP, eds: *Functional Investigation in Esophageal Disease.* Basel, Karger, 1994, p. 71.

14. Klauser AG, Schindlbeck NE, Mueller-Lissner SA: Esophageal 24-hour pH monitoring: Is prior manometry necessary for correct positioning of the electrode? *Am J Gastroenterol* 85:1463, 1990.

15. Mattox HE, Richter JE, Sinclair JW, Price JE, Case LD: Gastroesophageal pH step-up inaccurately locates proximal border of lower esophageal sphincter. *Dig Dis Sci* 37:1185, 1992.

16. Anggiansah A, Bright N, McCullagh M, Sumboonanonda K, Owen WJ: Alternative method of positioning the pH probe for oesophageal pH monitoring. *Gut* 33:111, 1992.

17. Singh S, Price JE, Richter JE: The LES locator: Accurate placement of an electrode for 24 hour pH measurement with a combined solid-state pressure transducer. *Am J Gastroenterol* 87:967, 1992.

18. Johnson LF, De Meester TR: Development of a 24 hour intraesophageal pH monitoring composite scoring system. *J Clin Gastroenterol* 8(suppl. 1):52, 1986.

19. Weiner GJ, Richter JE. Cooper JB: The symptom index: a clinically important parameter of ambulatory 24 hour esophageal pH monitoring. *Am J Gastroenterol* 83:358, 1988.

20. Emde C, Gardener A, Blum AL: Technical aspects of intralumenal pH-metry in man. Current status and recommendations. *Gut* 28:1177, 1987.

# 3

# The Performance and Interpretation of Ambulatory Esophageal pH Monitoring in Adults

Jean Price, MT (ASCP)
Joel E. Richter, M.D.

How do you actually set up an ambulatory esophageal pH study? The person who places the calibrated equipment and gives instructions to the patient needs to understand why the test was ordered and what useful information will be gleaned from this procedure. What specific symptoms are the focus of the study? This information must be communicated to the patient. If the patient understands what is expected and sees a logical rationale for the test, a more representative or typical day (as best a patient can have with a tube in his or her nose) will be monitored. The patient should strive for a normal daily routine and specifically perform activities that, in the past, have produced symptoms or events for which the patient is to be evaluated.

## INDICATIONS FOR AMBULATORY ESOPHAGEAL pH MONITORING

The purpose of long-term esophageal pH monitoring is to determine the answer to one or sometimes two questions. First, is acid present in the esophagus for an abnormal amount of time? It is known that brief acid reflux episodes occur in almost everyone, especially after meals. This is referred to as physiologic reflux. Therefore, if the pH study shows acid reflux occurring for a percentage of time that is greater than normal physiologic reflux, the subject is considered to have an abnormal amount of acid contact time. Second, esophageal pH monitoring is helpful for determining whether the occurrence of a specific symptom (e.g., chest pain, wheezing) is associated with a period of acid reflux. This relationship is valuable in determining the source and possible treatment of the patient's

**TABLE 3.1. INDICATIONS FOR AMBULATORY pH MONITORING**

| Gastroesophageal Reflux Disease |
| --- |

Typical Symptoms That Do Not Respond to Therapy:
   Heartburn
   Regurgitation
Atypical Symptoms:
   Non-cardiac chest pain: cardiac source must be ruled out first
   ENT Symptoms; hoarseness, laryngitis, globus, subglottic stenosis
   Pulmonary symptoms; asthma, chronic cough, aspiration pneumonia
Antireflux surgery
   Before—to document abnormal reflux
   After—if no relief from symptoms or to document arid control

symptoms. A positive correlation between symptoms and episodes of acid reflux may, in some instances, occur without overall abnormal esophageal acid contact time.

The patients who are referred for ambulatory esophageal pH monitoring may have a wide variety of problems and symptoms (Table 3.1). In general, this study is not routinely performed on patients with typical symptoms of gastroesophageal reflux disease who respond to traditional therapy. If, however, a patient has atypical symptoms or does not improve with traditional antireflux therapy, as esophageal pH monitoring study may be indicated.

Chest pain mimicking cardiac angina may be an atypical presentation of acid reflux disease. After a cardiac source of symptoms has been carefully excluded, an ambulatory esophageal pH study may be a reasonable next step.

An otolaryngologist may refer patients for pH monitoring who have symptoms of laryngitis, hoarseness, globus, or subglottic stenosis. Likewise, atypical pulmonary presentations of acid reflux include asthma, aspiration pneumonia, or chronic cough.

Prolonged pH monitoring may be helpful in patients whose symptoms are intractable to standard medical therapy. Studying these patients during treatment may identify those who need more acid suppression (who still have abnormal pH parameters) or those whose symptoms may not be related to acid reflux (who have normal pH parameters despite poor symptom control).

Prior to performing antireflux surgery, surgeons should document the presence of abnormal esophageal acid exposure by means of pH monitoring, especially in a patient whose symptoms are ambiguous or atypical. Finally, pH monitoring may be helpful after "seemingly successful" antireflux surgery to document the reason for persistent symptoms and the adequacy of arid control.

# PATIENT PREPARATION

At the time the study is scheduled, the patient should be informed that the test usually can be done on a outpatient basis. Since the subject will not be sedated, he or she can drive to the appointment and return home without difficulty after the equipment has been placed. The patient should not ingest any food and should limit liquid intake to a minimal amount for 6 hours prior to the study. An empty stomach helps decrease the likelihood of vomiting during placement of the probe and limits the buffering of gastric secretions, thereby usually making it easier to verify and assess gastric acidity.

## Appropriate Clothing

The patient should wear a shirt or blouse that buttons in the front to facilitate placement of the external skin reference electrode on the chest. Comfortable clothing that permits a normal activity level and be easily removed at bedtime is important in helping to provide an adequate, representative study period.

## Medications

If the pH study is being performed to diagnose gastroesophageal reflux disease, all antireflux medications should be discontinued. Histamine $H_2$ receptor antagonists and promotility drugs should be stopped a minimum of 48 hours prior to the study time. Proton pump inhibitors should be stopped at least 7 days prior to the study. Antacids may be permitted up to 6 hours prior to probe placement. Generally, other medications should be continued if they are part of the patient's normal daily routine. This decision should be individualized by the physician ordering the test on a case by case basis. Increasingly, the pH study is performed to assess effectiveness of medical antireflux regimens. In this case, all antireflux medications should be continued.

## Scheduling

Often a patient is scheduled to have multiple procedure and diagnostic evaluations in a short period of time. Can a pH test be done in conjunction with these and, if so, what is an appropriate order for the procedures?

### ENDOSCOPY

Same day endoscopy potentially can interfere in obtaining a reliable pH study. The sedative medication given for an endoscopic examination may distort the motility study. Often the patient may not remember the diary instructions for the pH test after being sedated. Also, the patient may be sleepy the rest of the day and not be able to perform his normal activities. If there is no other alternative, do the endoscopy in the early morning and the pH evaluation in the late morning or early afternoon. Alternatively, the endoscopy may be done the day the patient returns to have the pH probe removed.

### BARIUM SWALLOW

If a barium swallow is being performed first, wait 1–2 hours before starting the pH test so the barium has time to empty from the stomach. This should prevent the patient from vomiting the barium and also the buffering of the gastric contents.

### Radionuclide Gastric or Esophageal Emptying

Wait 1–3 hours after the test for the contents to pass.

### Pulmonary Function Test

Check with your pulmonary specialist. Technically, this test can be done with the pH probe in place.

### Other Radiographic Studies

The pH monitor must be shielded. If it is exposed to X-ray beams, the data may be erased.

# TIME INVOLVED AND LOCATION OF THE SPHINCTERS

Before beginning the actual pH test, a brief history of the patient's symptoms should be taken to help clarify the purpose and focus of the study. A simple description of the test procedure and what to expect during the study are helpful. A photograph of a subject wearing the pH equipment may be helpful and reassuring to the patient.

If the location of the lower esophageal sphincter (LES) has already been determined manometrically, allow 30 to 45 minutes for placement of the pH equipment. If a previous manometric reading has not been obtained, a brief study should be done to locate the LES and also the upper esophageal sphincter (UES), if a dual probe is to be used. Doing a manometric study increases the time of the study by approximately 30 minutes.

When performing an esophageal manometry study to locate the sphincters, an abbreviated version of the standard manometry study may be performed. The manometry catheter is passed nasally (like the pH probe) and advanced into the stomach. A station pull-through of the lower esophageal sphincter is performed by slowly withdrawing the catheter from the stomach, through the LES, and into the esophagus, in 0.5 cm increments. The proximal border of the LES is identified as the high pressure area directly before entering the body of the esophagus. The motility of the esophageal body is next assessed by giving and observing five to seven wet swallows (5 ml). If these swallows appear abnormal, a full 15 wet swallows should be given and observed. Finally, the catheter is withdrawn until the location of the UES is identified. As in the lower sphincter, a slow station pull-through is performed, moving the catheter at increments of 1 cm. The location of the distal and proximal borders should be determined. Different institutions will place the proximal sensor just below the distal border of the UES, in the UES, or above the proximal border of the UES. (See Chapters 12 and 13). Both techniques require the availability of dual probes with variable lengths separating the sensors. Another more practical approach is to use a standard dual pH probe (15 cm separation between the sensors), placing the distal sensor 5 cm and the proximal sensor 20 cm above the LES.

# pH CALIBRATION

Use a fresh battery for each test. Insert the battery into the recorder just before calibration because the voltage will begin to decrease even when not recording. Clear the previously stored data from the recorder. Use only buffer solutions recommended by the manufacturer. Follow storage instructions and observe the expiration dates. Gently mix the buffers before use.

pH probes are available with either internal or external reference electrodes. Internal reference probes are single use only. External reference probes can be reusable or are semi-disposable. Calibration methods vary depending on the manufacturer of the recorder and on the type of probe used. Follow the manufacturer's instructions carefully. If using a dual pH probe, make sure both sensors are immersed in the buffer solution during calibration.

Internal reference electrodes have a protective sleeve that should be removed just prior to calibration. The probe then must be soaked in buffer to activate the reference electrode. When calibrating the probe, make sure that the buffer solution covers the reference electrode.

External reference probes have either an attached reference electrode or use a silver-silver chloride skin electrode. When calibrating with an attached reference electrode, both the pH probe and the reference electrode should be placed in the buffer solution. When calibrating with the skin electrode, the electrode should be placed on the patient's chest. The pH probe and one of the patient's fingers should be placed in the buffer solutions during calibration.

# PLACEMENT OF THE pH PROBE

Assemble all the necessary supplies. Items needed include tissues, a cup of water with a straw, precut tape, an emesis basin, surgical lubricating jelly, gauze pads or alcohol wipes, topical anesthetic, gloves, and the pH equipment (Figure 3.1). Have the patient sit either on a stretcher or in a chair.

If using a pH probe with an external electrode, place it on the chest over an area that remains the same shape and conformation in different positions, such as on the manubrium sterni. Excessive hair should be removed with a razor. Clean the area well with an alcohol wipe or a moistened gauze pad with soap. Remove the soap with another moistened gauze pad. This step is very important because poor electrode adherence will produce a faulty electrical circuit, resulting in artifacts; thereby making the study partially unusable or at the worst, entirely uninterpretable.

By convention, the distal probe will be placed 5 cm above the proximal border of the LES. The location of the proximal electrode has not been standardized. Some institutions place the sensor 1 to 2 cm below the distal border of the UES, whereas others place the probe 2 cm above the proximal border of the UES. If using a pH probe that does not have markings to indicate length, measure the probe and mark it using a marker.

Always wear gloves when inserting the probe. Have the patient remove his or her glasses. If dentures or partial plates are securely in place, they do not have to be removed. Explain to the patient how the tube will be inserted and what they should do to help in this process. One will feel pressure as the probe is passed through the naris. Once the probe enters the pharynx, the patient may experience gagging or coughing. Reassure the

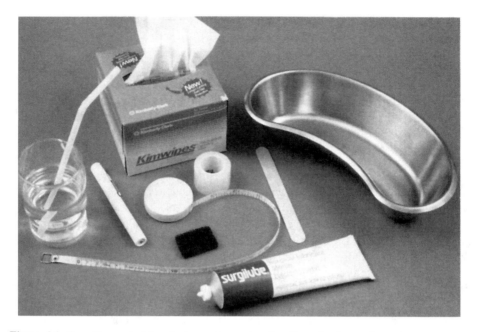

**Figure 3.1.** Supplies needed for placing and securing the pH probe. Items shown include tissues, emesis basin, glass of water with straw, penlight to ensure proper probe location, tape measure, precut tape, tongue blade, and surgical lubricating jelly.

patient that he/she will be able to breath normally during this process. Usually it is preferable to place the probe on the side opposite the patient's dominant hand so the probe does not get in the way during meals. If there is an anatomic obstruction or the patient requests that you do so, the probe may be inserted on the dominant side. With the patient sitting and head level, insert the lubricated pH probe into the naris and pass the probe straight back. When the probe enters the pharynx, have the patient bend the head forward, as if praying. This maneuver will guide the probe into the esophagus and not into the trachea. Have the patient begin taking small sips of water to facilitate passage of the probe. Pass the probe into the stomach and note the pH. Gastric pH should be below 4.0 unless the patient is on acid suppressive medication. If the pH does not drop below 4.0, the correct placement of the probe should be verified in one of two ways. The most accurate method is to fluoroscope or X-ray the patient. The other way is to have the patient take a small sip of fruit juice or soda. If the probe is positioned correctly in the esophagus, it will take several seconds for the pH to drop. If it is curled in the esophagus, the tip will be in the proximal esophagus or pharynx, and the pH will drop very quickly. Remove the probe and insert it again. If the pH does not drop, the probe may be in the trachea or lungs. Usually, if this occurs, the patient will experience coughing. Remove the probe and reinsert.

Once proper placement of the probe has been verified, pull the probe back so that the distal sensor is 5 cm above the proximal border of the LES as previously determined by manometry. Tape the probe securely to the nose. Secure the probe to the cheek and then loop the probe over the ear and tape it on the neck (Figure 3.2). Turn on the recorder. Place the recorder in a carrying case and attach to a belt or shoulder strap. To avoid catching the wires on objects, place the wires behind the clothing.

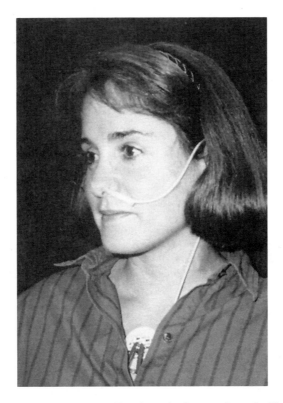

**Figure 3.2.** Appropriate placement of the pH probe and reference electrode. The pH probe is taped securely near the nose, avoiding unnecessary tension on the edge of the nose. The probe is looped over the ear with additional taping to the cheek and lower neck. The reference electrode is placed on the chest over an area that does not change shape with movement, such as the manubrium. In this position, the pH probe allows full range of movement and usually is not obtrusive.

# DIARY KEEPING AND EVENT MARKING

Reassure the patient that he/she will be able to tolerate the pH probe with little discomfort. Stress the importance of maintaining a normal level of activity and of eating normally as possible. We do not routinely tell the patient how to remove the equipment, but do provide telephone numbers and instructions to call if one feels the probe must be removed. Patients who have major complaints of vomiting will occasionally vomit up the probe. We do instruct these patients on how to remove the probe if this happens.

Some physicians request a nonacidic diet containing foods with a pH of 5.0 or greater. Others prefer to have the patient eat ad libitum, and any type of food is permissible. The latter approach is now more popular as it allows better assessment of symptoms relative to normal activity at home or work. Any beverage is allowed with meals and snacks. The only beverage allowed between meals and snacks is water. If a person normally drinks alcohol and/or smokes tobacco, these habits should be continued. Ask the patient not to chew gum or eat hard candy, as these will increase saliva production and can affect the results of the test. Instruct the patient to take only a sponge bath, not a shower or bath as these could get the equipment wet.

Most monitors have markers to record eating, sleeping, and symptom events. A written diary may also be used along with the markers. The patient should be instructed to record the beginning and end of each meal or snack and also the beginning and end of any period where one is supine, i.e., nap or bedtime. Correlation of symptoms with acid reflux is an important part of the test. Instruct the patient to accurately record the onset of any symptoms experienced during the study.

# PROBLEMS TO ANTICIPATE

## Eating

During eating, the probe may move slightly during swallowing. The patient should be instructed to eat slowly, chew thoroughly, and drink plenty of liquids. Stress the importance of eating normal meals. If the patient eats very little, a false negative study may result.

## Rhinorrhea

Patients may experience some watering of the eyes and nasal dripping. This should decrease over time, usually within an hour. The patient should be instructed to gently blow the nose. If this does not clear up the rhinorrhea, antihistamines or decongestants may be taken.

## Lump in the Throat

The patient will feel the probe slightly during the study. Reassure the patient that the probe is well secured and will not come out under normal conditions.

## Emergency Removal of Equipment

Rarely, the patient may vomit up the probe. The probe will be visible in the patient's mouth. Instruct the patient to untape the probe, take a deep breath, and then pull out the probe. The patient should wrap the probe in a plastic bag or towel and return to the laboratory at the scheduled time. When the patient returns, turn off the monitor and download as usual. The analysis can be edited to remove the time while the probe was removed.

## Failure to Return

Occasionally, a patient cannot return because of illness or inclement weather. If in the standby mode, most recorders can retain the data for several days. Have the patient return as soon as possible and download the recording.

Rarely, a patient will remove the equipment and intentionally not return. In some large urban areas, medical devices such as Holter monitors can be traded on the street or pawned for cash. As part of your regular informed consent, you might consider having a clause in which the patient assumes financial liability for loss of the equipment. However, our lab has never had a problem with over 3000 studies during the last seven years and does not require the patient to sign a liability form.

# TERMINATION OF THE STUDY

The monitoring period does not have to be a full 24 hours. A recording of 18 to 24 hours is adequate in most studies (see Chapter 5). The study should include two to three meals and a supine period.

When the patient returns, check to see if the equipment is functioning properly while it is still running. Turn off the recorder. Untape the probe, have the patient take a deep breath and hold, then remove the probe from the nose. Check the placement of the pH probe as soon as it is removed. Rarely, if not taped securely, the probe can slip during the study, or it may have been taped in the wrong place. Any change in placement should be documented, and the results should be interpreted accordingly. Remove the reference electrode if necessary.

Some manufacturers recommend post calibration of the probe. Follow their instructions for the post calibration procedure. The values will not be exactly the same as the precalibration values, but should be within 0.5 pH units of the original values.

Review the diary carefully with the patient. Download the recording to a computer. Carefully review the tracing on the computer screen or printout. The tracing should be viewed on 1 or 2 hour plots, not the compressed 24 hour plot. Symptom correlation should be calculated.

# DISINFECTION

Single use probes should be discarded after using on one patient. Reusable and semi-disposable pH probes should be disinfected after each use. The probes are to be cleaned first using an enzymatic cleaner. Cold gas sterilization or high level disinfectant solutions, such as glutaraldehyde, may be used. Recommended soaking times vary depending on the solution used. Follow the manufacturer's instructions carefully.

# COMMON TECHNICAL PROBLEMS
# IN INTERPRETING pH STUDIES

Occasionally, problems will occur during a study that require portions of the study to be deleted from the study. If the electrode becomes disconnected, the recording drops suddenly to zero (Figure 3.3). Rarely, a probe will fail during the study. The reading may suddenly drop to zero, or it may slowly drift towards zero (Figure 3.4). When using a dual channel pH probe, the proximal sensor may lose contact with the esophageal wall and "dry out," especially during sleep when the patient does not swallow. When this occurs,

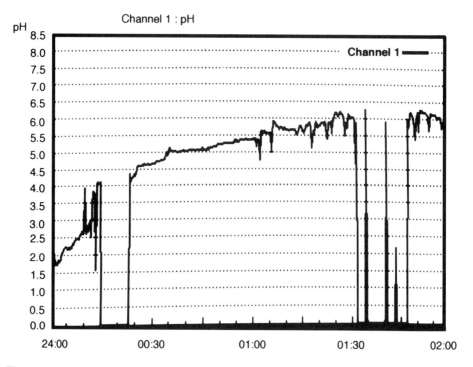

**Figure 3.3.** Technical problem: External reference electrode is intermittently becoming disconnected at the recorder or possibly poor skin contact. Note the pH recording drops suddenly to zero. These portions need to be excluded from the computer analysis.

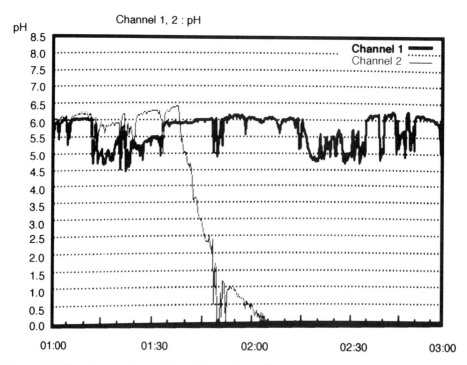

**Figure 3.4.** Technical problem: Probe failure. The reading may suddenly drop to zero or it may slowly drift towards zero.

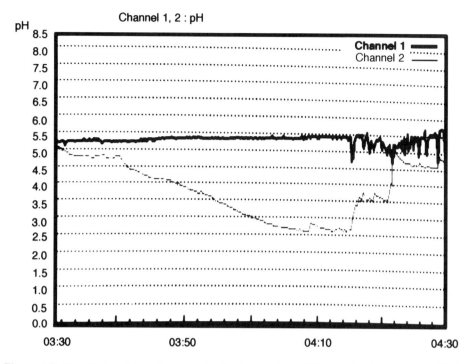

**Figure 3.5.** Technical problem: "pseudoreflux". The proximal pH sensor has lost contact with the esophageal wall and "drys out." When this occurs, the proximal reading slowly drifts down then abruptly returns back to baseline when contact with the esophageal mucosa is restored. Beware: the computer will analyze this artifact as a reflux episode. This must be excluded from the analysis.

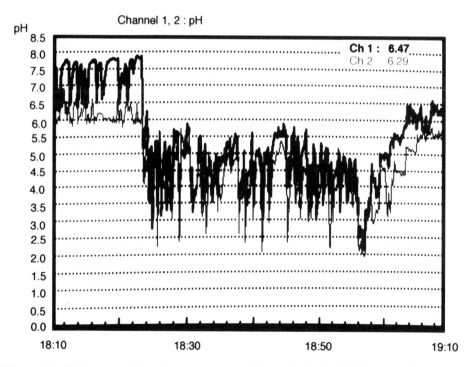

**Figure 3.6.** "Reflux" episodes related to a long meal from 18:15 to 19:00. We normally do not exclude meals routinely unless they interfere with study interpretation. This emphasizes the importance of carefully reviewing all aspects of the pH training before the computer analyzes the data.

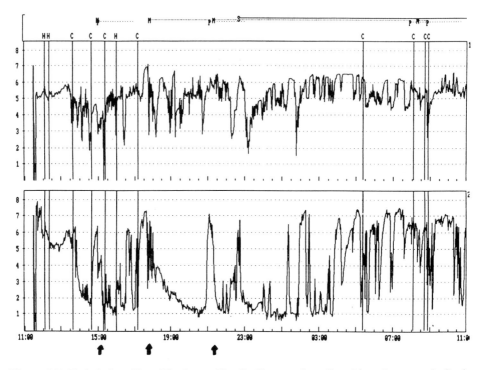

**Figure 3.7.** Technical problem. The lower (distal) pH sensor has slipped into the stomach. Study from 1400 to 0500 needs to be deleted. Characteristics suggesting there is not prolonged esophageal reflux: (1) esophageal pH infrequently less than 2, and (2) distal pH rises, rather than falls, in response to meals (dark arrows) as the result of gastric acid neutralization by foods.

the proximal reading slowly drifts down, then will suddenly return to a normal pH reading when the patient swallows (Figure 3.5). When calculating the amount of acid reflux, periods such as these need to deleted from the calculations. Failure to do so may result in falsely elevated proximal results.

Another problem may occur if the patient consumes very acidic meals or beverages. In most cases, the amount of time the pH is less than 4.0 due to meals is negligible. However, if the amount of time is prolonged, the meals can be deleted from the calculations (Figure 3.6). Finally, the distal pH sensor occasionally will slip into the stomach (Figure 3.7), but the degree of acidity (pH < 2) and relationship of meals to pH changes will give important clues to this technical problem.

# 4

# The Placement of the pH Electrode

Huitt E. Mattox, III, M.D.

The pH electrode for prolonged ambulatory intraesophageal pH monitoring is conventionally placed 5 cm above the gastroesophageal junction (GEJ). Since the lower esophageal sphincter (LES) normally coincides with the GEJ, the pH electrode usually has been positioned 5 cm above the proximal LES border. This placement requires esophageal manometry to locate the sphincter prior to pH monitoring. However, manometry otherwise provides limited information about acid reflux and is not available to many physicians. Therefore, other methods of GEJ localization, such as endoscopy, radiography, and the pH change between the stomach and the esophagus, have been advocated as alternatives to manometry for pH probe placement. In fact, the method used for localizing the GEJ really does not matter as long as the technique is accurate and reproducible from patient to patient. With this in mind, three points will be addressed: (1) the importance of accurate GEJ localization, (2) alternative methods of GEJ localization and their relative advantages and limitations, and (3) the reasons why pH probe placement based on locating the LES by manometry is superior to other methods and should remain the gold standard.

## IMPORTANCE OF pH PROBE LOCATION

Traditionally, an episode of acid reflux has been defined as a drop in the intraesophageal pH to 4.0 or below, as detected by a probe placed 5 cm above the GEJ. There is nothing magical about the 5 cm distance. It was chosen as a point clearly within the esophageal body that was close enough to the GEJ to detect reflux but not too close so as to cause the probe to advance into the stomach when the esophagus shortened with swallowing.

Recent studies have shown that the pH electrode position within the esophagus has a marked effect on quantitating the amount of acid reflux. In 24 normal volunteers, Lehman and colleagues reported that nearly twice as many reflux events were detected by a probe placed 1 cm compared with 5 cm above the GEJ.[1] These differences were not "false reflux" resulting from the distal probe advancing into the stomach because prior radiographic studies had demonstrated that the pH probes did not change with altering body positions.[2] Other studies in adults and children have shown that significantly less reflux is detected in the proximal as compared with the distal esophagus.[3-5] For example, Johansson and Tibbling found that more reflux was detected in 39 reflux patients by a probe placed 5 cm compared with 15 cm above the GEJ ($9.0 \pm 7.9$ percent versus $2.0 \pm 2.0$ percent).[3] Regional differences were probably attributable to gravity, esophageal peristalsis, and salivary secretions that either prevented or diluted reflux into the proximal esophagus.

These studies have convincingly shown that varying the position of the pH probe within the esophagus directly affects the amount of acid reflux detected. Therefore, if the position of the GEJ reference point is imprecise, or worse if it is erroneous, the true distance between the GEJ and the probe will not be known in any given patient and will likely vary from patient to patient. For example, a patient with normal reflux parameters may be inappropriately considered abnormal if recordings were made with the probe positioned less than 5 cm from the GEJ. On the other hand, a patient with excessive esophageal acid exposure may be considered normal if recordings were made from a probe positioned in the more proximal esophagus. These problems are compounded by the inability to interpret individual test results unless the alternative technique has been shown to be as accurate as manometric localization because current "normal" pH values are only applicable if the GEJ to pH probe distance is 5 cm.[6-9]

# ALTERNATIVE METHODS FOR pH ELECTRODE PLACEMENT

## Upper Gastrointestinal Series or Endoscopy

The requirements of pH monitoring place special restrictions on alternative techniques for localizing the GEJ prior to probe placement. Since the pH probe is positioned relative to the anterior nares, the GEJ needs to be recorded in terms of the distance from the anterior nares, and the length and contour of the nasopharynx should be considered. Commonly used tests such as upper gastrointestinal series or upper endoscopy routinely assess the GEJ but do not evaluate the nasopharynx and are unable to accurately provide the distance between the anterior nares and the GEJ. Consequently, there are no studies attempting to determine this distance using upper gastrointestinal barium studies. When the localization of the GEJ by gastroscopy and manometry were compared, Walther and DeMeester found a poor correlation ($r = 0.54$, $p < 0.02$) between the techniques and wide variability[10] (Figure 4.1). Therefore, the ideal alternative method for localizing the GEJ would need to include the dimensions of the nasopharynx and preferably not require additional equipment or expense other than that needed for pH monitoring.

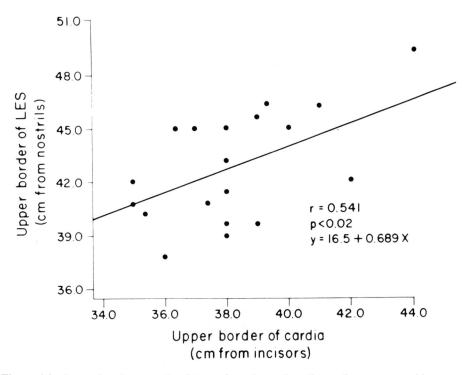

**Figure 4.1.** Comparison between the distance from the teeth to the cardia, as measured by upper gastrointestinal endoscopy, and the distance between the nostrils and the upper border of the LES, as measured by manometry. (From Walther B, DeMeester T: Placement of the esophageal pH electrode for 24-hour esophageal pH monitoring. In: DeMeester TR, Skinner DB, eds: *Esophageal Disorders: Pathophysiology and Therapy.* New York: Raven Press, 1985:539–541, with permission.)

## FLUOROSCOPY

The internal electrical wiring of the pH probe is radiopaque, thus enabling it to be imaged fluoroscopically. Assuming that the GEJ coincides with the diaphragmatic hiatus, the probe could be fluoroscopically placed 5 cm above the diaphragm with the reasonable expectation that it would be 5 cm above the GEJ. Unfortunately, this relationship is only true in healthy adults; hiatus hernias, common in those with gastroesophageal reflux disease displace the GEJ proximal to its typical location. When Klauser and coworkers compared manometric and fluoroscopic GEJ localization in 74 subjects, there were large differences between the two techniques (Figure 4.2). As might have been expected, fluoroscopy had the poorest correlation with manometry in patients with hiatus hernia.[11] Furthermore, the logistical problems of moving a patient from the gastrointestinal laboratory to the fluoroscopy suite are difficult in most hospitals and impractical in a clinic setting.

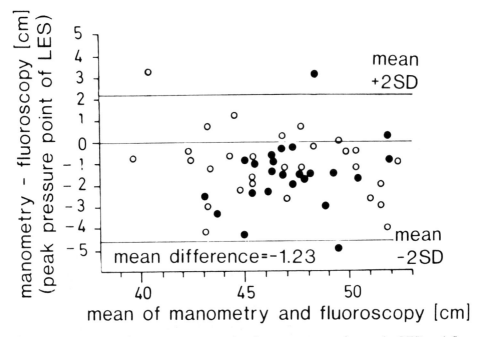

**Figure 4.2.** Agreement between manometry (maximum pressure point on the LES) and fluoroscopy for localizing the GEJ. The zero line represents perfect agreement. The difference between each of the two values measured by the respective two methods is plotted against their mean by manometry and fluoroscopy. This analysis assumes that the location of the GEJ is not precisely known but is best estimated by averaging values obtained by each method. ○ Normal pH test ($N = 29$). ● Pathologic reflux on pH testing ($N = 24$). (From Klauser A, Schindlbeck N, Muller-Lissner S: Esophageal 24-h pH monitoring: It prior manometry necessary for correct positioning of the electrode? *Am J Gastroenterol* 85:1463–1467, 1990, with permission.)

## pH CHANGE

The pH change, somewhat like the manometric LES, is a physiologic rather than an anatomic correlate of the GEJ. The pH change occurs relatively abruptly and can be measured either by withdrawing a pH electrode from the stomach into the esophagus or by advancing the electrode from the esophagus into the stomach.[12] Therefore, it has been advocated as an acceptable alternative to manometry prior to probe placement.[11] The attractiveness of this technique is that it can be detected with currently available pH equipment, does not add expense to the procedure, and avoids the patient's being intubated by the manometry catheter. However, there is considerable debate over the accuracy of the pH change compared with the accuracy of manometry in locating the GEJ.

Two studies, each with potential methodologic problems, have concluded that the pH change does not correlate with the manometrically identified LES in a majority of reflux patients and healthy volunteers (N = 52).[10,13] Walther and DeMeester reported that the pH change could only identify the proximal LES within ±1 cm in 25 percent of subjects. The pH change was determined by pushing the electrode from the esophagus into the stomach because previous work had shown the pH gradient to be shorter with antegrade versus retrograde probe movement.[12] Unfortunately, this may have caused the pH change to overestimate the GEJ because of buckling of the probe with advancement.

Marples and colleagues, utilizing a more slender pH probe, pulled the electrode from the stomach into the esophagus, thereby avoiding problems with buckling.[13] They concluded that asymptomatic volunteers would have had the electrode placed, on the average, 6 cm below (caudad) and reflux patients 2 cm above (cephalad) the ideal manometric position if the pH probe had been positioned relative to the pH change. The presence of acid in the distal esophagus could have accounted for the more cephalad pH change in reflux patients. In contrast, the pH change could have overestimated the GEJ in asymptomatic volunteers because the patients were sitting when the probe was retracted. In the upright position, acid tends to pool in the more distal stomach, causing the pH change to occur caudad to the GEJ and thereby overestimating the distance of the anterior nares from the GEJ.

More recently, a West German group and our laboratory have independently re-evaluated the accuracy of the pH step-up for GEJ localization.[11,14] Both groups studied subjects in the supine position, choosing to retract rather than advance the probe because of the

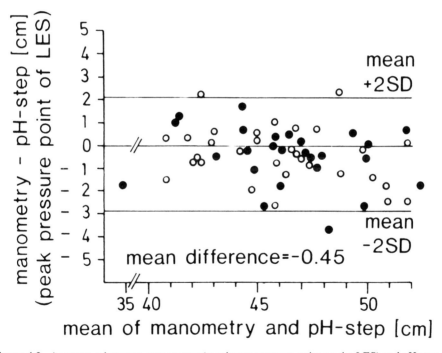

**Figure 4.3.** Agreement between manometry (maximum pressure point on the LES) and pH step-up for localizing the GEJ. The difference between each of the two values measured by the respective two methods is plotted against their mean. A similar analysis using the proximal LES border as the manometric reference found a mean difference of −1.31 ○ Normal pH test (N = 31). ● Pathologic reflux on pH testing (N = 26). (From Klauser A, Schindlbeck N, Muller-Lissner S: Esophageal 24-hr pH monitoring: Is prior manometry necessary for correct positioning of the electrode? *Am J Gastroenterol* 85:1463–1467, 1990, with permission.)

flimsiness of modern pH electrodes. In 60 patients and 14 asymptomatic volunteers, Klauser and associates showed relatively small differences ($-0.45$) between manometry and the pH change, as assessed by the statistical method of Bland and Altman.[11,15] Using this analysis, it is assumed that the true location of the GEJ is not known and must be determined by indirect methods. They further assumed that the GEJ probably lies somewhere between the pH change and the LES. Therefore, the difference between manometry and the pH step-up for localization of the GEJ was compared with the mean values of manometry and pH step-up (Figure 4.3).

There are several aspects of Klauser and colleagues' findings that can be challenged. Foremost, it may be inappropriate to assume that the GEJ lies "between the pH change and the manometric LES," since the location of the pH change may be altered by reflux episodes. It is also of concern that the investigators did not report their raw data, which would show the absolute differences between individual pH step-up and manometry values. This is an important omission because the interpretation of pH test results may be confounded in individual cases if there is excessive disagreement between the pH change and the findings manometry. This is based on the fact that the detection of acid reflux events is directly influenced by the position of the probe within the esophagus. Finally, the investigators formulated their conclusions by comparing the pH change to the "maximal pressure point" of the LES, which typically occurs in the midportion of the LES just prior to the respiratory inversion point.[16] However, current normal pH values assume that the probe is positioned 5 cm above the proximal LES rather than at the maximal pressure

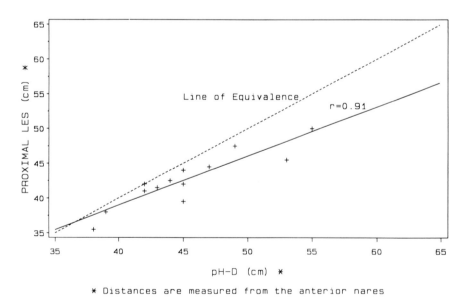

**Figure 4.4.** Manometric location of the proximal LES border versus the pH step-up in 14 asymptomatic volunteers. The solid line (-) is the regression line for the volunteers and is shown in comparison to the line of equivalence (---). The pH step-up tended to correlate better with manometry in asymptomatic volunteers compared within patients ($p = 0.061$).

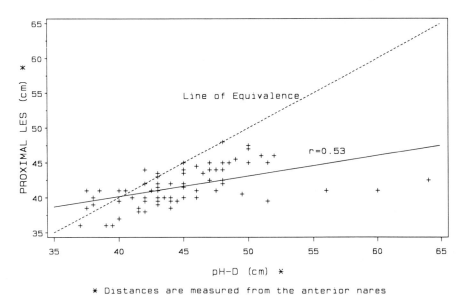

**Figure 4.5.** Manometric location of the proximal LES border versus the pH step-up in 71 symptomatic patients. The solid line (-) is the regression line for the patients and is shown in comparison to the line of equivalence (- - -). Although the correlation between the pH step-up and manometric LES localization was significant ($p < 0.0001$), the pH step-up overestimated the proximal LES border by an average of 3.3 cm.

point. In fact, when Klauser and coworkers compared the pH change with the position of the proximal LES, the pH change fared no better than results with fluoroscopy (mean difference $= -1.31$), which the authors had concluded was not an acceptable method of localizing the GEJ.

In our laboratory, we have studied the relationship between the pH step-up and the manometrically defined GEJ in 14 asymptomatic volunteers and 71 patients referred for pH monitoring. We used the proximal LES border for the comparison point because it is the conventional reference point from which standard normal pH values have been determined.[6-9] Regression lines derived from our subjects showed good correlation ($r = 0.91$) between the pH change and manometry in healthy volunteers (Figure 4.4). However, the correlation ($r = 0.53$) was relatively poor in our patients with gastroesophageal reflux disease (Figure 4.5). Further data analysis showed that the presence of abnormal reflux parameters or esophagitis significantly contributed to this error when locating the proximal border of the LES by the pH step-up. Obviously, this limitation is alarming because reflux patients are the most likely subjects to undergo pH testing. More important, Table 4.1 shows that the pH step-up could not accurately locate the proximal LES within $\pm 3$ cm (6 cm total span) in 58 percent of symptomatic patients and 29 percent of asymptomatic volunteers.

**TABLE 4.1. ACCURACY OF LOCALIZATION OF LOWER ESOPHAGEAL SPHINCTER BY pH STEP-UP**

| | | Absolute Difference Between pH Step-up and Proximal LES (cm) | | | | | |
|---|---|---|---|---|---|---|---|
| | | 0 | 0–1 | 1–2 | 2–3 | ≥3 | Total |
| Symptomatic patients | N(%) | 5(7) | 3(4) | 11(15) | 11(15) | 41(58) | 71(100) |
| Asymptomatic volunteers | N(%) | 1(7) | 0(0) | 7(50) | 2(14) | 4(29) | 14(100) |

# SUMMARY

Although manometry adds limited information about gastroesophageal reflux disease, the majority of studies have concluded that it is the only technique that accurately locates the GEJ. Furthermore, most currently accepted pH standards were established using manometry as the reference point for the pH electrode placement. Currently, several manufacturers have developed "LES locators" which combine a single solid state pressure transducer with a pH electrode. As shown in Figure 4.6, this technique accurately identifies the proximal LES border compared to perfused manometry and requires only a single intubation.[17]

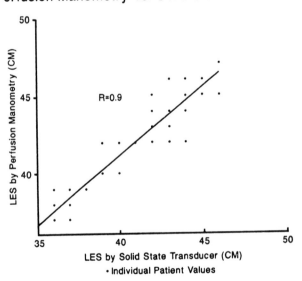

**Figure 4.6.** Linear regression analysis found a very strong correlation ($r = 0.9$) between the values locating the proximal LES border obtained by stationary perfusion manometry *versus* the solid state transducer.

# REFERENCES

1. Lehman G, Rogers D, Cravens E, et al: Prolonged pH probe testing less than 5 cm above the lower esophageal sphincter (LES). Establishing normal control values. *Gastroenterology* 98:77, 1990.

2. Lehman G, O'Connor K, Cravens E, et al: Does placement of pH probes less than 5 cm above the lower esophageal sphincter (LES) produce falsely positive gastroesophageal reflux? *Gastroenterology* 94:A255, 1988.

3. Johansson K, Tibbling L: Gastric secretion and reflux pattern in reflux esophagitis before and during ranitidine treatment. *Scand J Gastroenterol* 21:487–492, 1986.

4. Haase Gl, Ross M, Gance-Cleveland B, et al: Extended four-channel esophageal pH monitoring: The importance of acid reflux patterns at the middle and proximal levels. *J Pediatr Surg* 23:32–37, 1988.

5. Sandheimer J, Haase G: Simultaneous pH recordings from multiple esophageal sites in children with and without distal gastroesophageal reflux. *J Pediatr Gastroenterol Nutr* 7:46–51, 1988.

6. Mattox H, Richter J: Prolonged ambulatory esophageal pH monitoring in the evaluation of gastroesophageal reflux disease. *Am J Med* 89:345–356, 1990.

7. Rosen S, Pope C: Extended esophageal pH monitoring. An analysis of the literature and assessment of its role in the diagnosis and management of gastroesophageal reflux. *J Clin Gastroenterol* 11:260–270, 1989.

8. Johanson L, DeMeester T: Twenty-four hour pH monitoring of the distal esophagus. *Am J Gastroenterol* 62:325–332, 1974.

9. Richter JE, Bradley LA, DeMeester TR, Wu WC: Normal 24-hour ambulatory esophageal pH values: Influence of study center, pH electrode, age and gender. *Dig Dis Sci*, 37:849–56, 1992.

10. Walther B, DeMeester T: Placement of the esophageal pH electrode for 24-hour esophageal pH monitoring. In DeMeester TR, Skinner DB, eds: *Esophageal Disorders: Pathophysiology and Therapy*. New York, Raven Press, 1985, pp. 539–541.

11. Klauser A, Schindlbeck N, Muller-Lissner S: Esophageal 24-h pH monitoring: Is prior manometry necessary for correct positioning of the electrode? *Am J Gastroenterol* 85:1463–1467, 1990.

12. Thurer R, DeMeester T, Lawrence F: The distal esophageal sphincter and its relationship to gastro-esophageal reflux. *J Surg Res* 16:418–423, 1974.

13. Marples M, Mughal M, Bancewicz J: Can an esophageal pH electrode be accurately positioned without manometry? In: Stewart J, Holsches A, eds: *Diseases of the Esophagus*. Berlin, Heidelberg, New York, Springer-Verlag, 1987, pp. 789–791.

14. Mattox H, Sinclair J, Price J, et al: The gastroesophageal (GE) pH step-up inaccurately locates the proximal lower esophageal sphincter *Dig Dis Sci* 1992;37:1185–91.

15. Bland J, Altman D: Statistical methods for assessing agreement between two methods of clinical measurement. *Lancet* 1:307–310, 1986.

16. Mattox H, Sinclair J, Price J, et al: The gastroesophageal pH step-up occurs at the mid-portion of the lower esophageal sphincter. *Gastroenterology* 98:A8, 1990.

17. Singh S, Price JE, Richter JE: The LES locator: accurate placement of an electrode for 24-hour pH measurement with a combined solid state pressure electrode. *Am J Gastroenterol* 87:967–976 1992.

# 5

# What are the Optimal Length and Setting for Esophageal pH Testing?

Cathy A. Schan, PA-C

As with many medical procedures, the variations in esophageal pH testing have been extensive. To coincide with specific needs and limitations of the medical staff and patients, these studies have been conducted over time periods as short as 30 minutes and up to 24 hours. The location of the study also has alternated between inpatient and outpatient settings. The goal with esophageal pH testing, as with any other medical test, is to find the most accurate time frame and location to produce interpretable data for diagnosis while maintaining a practical setting. This chapter reviews the available data on the optimal length and setting for esophageal pH testing.

## ESOPHAGEAL pH TESTING: 24 HOURS VERSUS SHORTER STUDIES

In quest of an even more convenient tool, researchers have asked if esophageal pH testing could be conducted over shorter periods of time. They perceived that studies lasting 24 hours in duration were too constraining for both the patients and the medical staff.[1] Furthermore, they believed that patients could physically tolerate the intubation of the pH wire better over shorter periods while still preserving accuracy.[1,2] Other arguments cited include less wear and tear on the equipment if used over shorter intervals, which translates into less overall financial output for maintenance.[1,3] Finally, if the time period is 8 hours or less, the equipment can be supervised as opposed to risking damage when patients take the monitors home.[3]

In the determination of study length, the first and foremost question should be one of accuracy of results. Table 5.1 outlines the sensitivities of pH studies based on time frames

**TABLE 5.1. SENSITIVITY OF VARIOUS DURATIONS OF ESOPHAGEAL pH TESTING**

| Study | No. of Subjects | Study Duration | | | | | | Reference Standard |
|---|---|---|---|---|---|---|---|---|
| | | 24 hr | 16 hr | 12 hr | 10 hr | 8 hr | Postprandial | |
| Fink/McCallum[1] | 16 patients<br>8 controls | 100% | — | 94% | — | — | 77% | 24 hr pH |
| MacMahon[3] | 155 patients | 100% | — | — | — | — | 85% | 24 hr pH |
| Jorgensen[4] | 59 patients | — | — | — | — | — | 85% | EGD* and histology |
| Klauser[5] | 289 patients | 93% | — | — | — | — | — | EGD |
| Schlesinger[6] | 34 patients<br>20 controls | 88% | — | — | — | — | — | EGD |
| Porro[7] | 20 patients<br>20 controls | 81% | — | 50% | 70% | — | — | EGD and histology |
| Fuchs/DeMeester[8] | 45 patients<br>45 controls | 96% | — | — | — | — | — | Patient history |
| Vitale[9] | 51 patients<br>22 controls | 76% | — | — | — | — | — | History |
| Johnsson[10] | 30 patients<br>30 normals | 87% | — | — | — | — | — | EGD |
| Walther/DeMeester[11] | 9 patients<br>11 normals | 100% | — | — | — | 100% | — | 24 hr pH |
| Grande[12] | 40 patients<br>15 controls | 100% | — | — | — | 97% | — | 24 hr pH |
| Choiniere[13] | 30 patients | 100% | — | — | — | 93% | — | 24 hr pH |
| Dobhan/Castell[14] | 43 patients | 100% | 86% | — | — | — | — | 24 hr pH |
| Dhiman[15] | 65 patients<br>16 controls | 86% | — | — | — | 79% | 54% | 24 hr pH |

*EGD = esophagogastroduodenoscopy.

ranging from limited postprandial periods to 24 hour studies. The reference standards for these studies vary from clinical history to endoscopy with or without histologic examination to 24 hour pH study because no perfect gold standard for studying gastroesophageal reflux disease exists. Nevertheless, this table clearly shows that 24 hour pH testing is the best tool for diagnosing acid reflux disease, with a sensitivity ranging from 76 to 100 percent. Perfect scores occur when pH monitoring is the reference standard, but when these studies are excluded, the 24 hour pH test still has excellent sensitivity, i.e., 76 to 96 percent.

Shorter testing periods have been advocated, including a 3 hour postprandial period, two consecutive 3 hour postprandial periods, and 8 to 16 hour periods usually divided between periods of monitoring with patients in the upright and supine positions. Postprandial monitoring has had variable sensitivity, depending upon the reference standards used. Several French groups[16] claimed that 3 hour pH monitoring after a standard meal had a sensitivity ranging from 86 to 91 percent when compared with manometric studies, scintigraphy, standard acid reflux test, and/or overnight pH testing. However, as shown in Table 5.1, post-prandial pH testing does not compare as favorably when 24 hour pH testing is used as the reference standard (77%–85%). If a shorter study period is desirable, 8 hour testing covering two meals and some time spent in the supine position may be a reasonably accurate alternative to 24 hour testing. Grande and colleagues[12] reported that combining the breakfast and lunch postprandial 3 hour periods gave a sensitivity of 97 percent and a specificity of 100 percent compared with 24 hour recording in distinguishing 15 controls and 40 patients with documented gastroesophageal reflux disease. Likewise, Walther and DeMeester[11] reported 100 percent sensitivity in 20 patients, whereas a French group found 93 percent sensitivity in 30 patients when 8 hour pH testing was compared with 24 hour testing.[13]

The reproducibility of any test is a measure of its accuracy. In a study by Johnsson and associates,[17] 20 patients underwent esophageal pH testing during two consecutive 24 hour periods. The results were analyzed, and comparisons were made between different time periods. They concluded that the reproducibility of pH testing increased as the length of the study increased. The 24 hour pH study period yielded a concordance rate of 83 percent, whereas the other time periods yielded less precise rates: 16 hours 58 percent, postprandial 11 percent, 6 hours 23 percent, and 4 hours 0 percent. Interestingly, the 12 hour period during this study had a reproducibility rate of 83 percent. Therefore, the authors concluded that 12 to 24 hours of esophageal pH testing had good reproducibility. Otherwise, limited studies involving only postprandial or night-time periods and studies lasting less than 10 hours generally had poor reproducibility[11–13,15]

A major problem with short postprandial studies is that they may miss abnormal reflux episodes that occur only when the patient is in the supine position. Although there is controversy over the importance of supine reflux,[18] we have seen 25 patients with abnormal acid reflux in the supine position only. Fifteen patients from this group had endoscopic examination, with six (40 percent) having evidence of esophagitis. Other groups support our findings with the observation that approximately 25 percent of patients with gastroesophageal reflux disease have abnormal amounts of acid reflux confined to supine periods.[9,15,19]

Another argument for conducting shorter studies is less wear and tear on the equipment and the need for less time on the part of medical personnel, which could then be translated into reduced costs for pH testing.[1,3] However, the pre-dominant factor in decreasing the longevity of the pH probes occurs during the disinfection process, which requires caustic agents. Regardless of the total length of the study, the duration of the disinfection phase remains constant. Furthermore, the medical personnel time is derived from interviewing and discussing the procedure with the patient, calibrating the equipment, intubating and removing the pH probe, and unloading the data into the computer. All of these times are fixed intervals regardless of the total time the patients wear the probes.

Patient tolerance is another rationalization for shorter studies.[2] Experience with intubation of the pH probe, both personal and observational, demonstrates that the most acute side effects occur within the first several hours. These side effects include irritation of the nasal cavity and throat, increased swallowing, and odynophagia. After the first several hours, the intensity of these side effects decreases, although they may persist in some patients. Since tolerance seems to vary widely from patient to patient, shorter studies may be preferable in patients who are particularly sensitive to the probe. However, these same patients may be diagnostic problems, and repeat testing will be difficult to obtain if the abbreviated study does not provide the diagnosis.

Porro and Pace[7] surveyed their patients regarding the test tolerance while wearing the pH probe for 10 hours, 12 hours, and 24 hours. Overall, patients admitted that shorter studies were more favorable than longer ones; however, the group as a whole did not express significant differences regarding tolerability. Ninety-five percent of patients stated that they were willing to repeat the 24 hour test. Interestingly, many of the complaints centered around the unattractiveness of wearing the probe as opposed to actual physical discomfort.

Besides the excellent sensitivity of the longer pH studies, there are several other advantages for extending the pH test for 18 to 24 hours. First, this length of time allows the practitioner to evaluate the circadium pattern of acid reflux, i.e., upright, supine, or combination pattern. The physician can then prescribe medication for gastroesophageal reflux disease that coincides with the specific pattern for that patient. In addition, the pattern can point out individual lifestyle habits contributing to reflux, such as supine reflux that is enhanced by late night snacking or upright reflux that is increased by cigarette smoking, gum chewing, or certain activities.

An important element of esophageal pH testing, especially with atypical presentations, is the symptom index.[5] The symptom index is a percentage derived by simply taking the number of symptoms coincident with reflux events and dividing by the total number of symptoms (see Chapter 9). Some patients may have normal total esophageal acid exposure times but have a "positive" symptom index, which informs the practitioner that a particular symptom is quite possibly secondary to esophageal acid exposure. By extending the duration of the analysis to 24 hours, the chance of the specific symptoms occurring is increased, and the symptom index can be utilized more fully for the total evaluation of the patient's problem.[14]

# INPATIENT VERSUS OUTPATIENT SETTING FOR ESOPHAGEAL pH TESTING

The first studies with esophageal pH testing were restricted by the available technology. During these early studies, patients had to stay overnight in the hospital while attached via cable to large, virtually immobile physiographs over a 24 hour period. Since those times, researchers have been looking for a simpler, more practical procedure. With the advent of lightweight, compact ambulatory monitors, esophageal pH testing was able to dispense with the necessity of admitting patients overnight to hospitals for the diagnosis of gastroesophageal reflux disease.

The ideal setting for performing the esophageal pH study is the environment that most closely represents the patient's daily physiologic state. Although the introduction of a foreign body does compromise patient activity to some degree, the outpatient setting comes the closest to normal day-to-day life. While away from the hospital, patients may encounter stresses related to their occupations or families. Also, patients feel free to continue any usual bad habits, such as over-eating of fatty foods and cigarette smoking, whereas strict supervision by the medical staff and family would be a hindrance to those who have a less than perfect lifestyle.

Hospitalization modifies two factors that normally contribute to acid reflux—food intake and activity. Food intake is modified in both quality and quantity. The typical patient eats less while in the hospital because of the inaccessibility of food at all hours and the undesirability of the food. Additionally, the hospital diet often has a lower total fat content than the patient's standard diet at home. Decreased food intake with a lower fat content results in less overall acid reflux because the amount of gastric distention is reduced.

When patients are hospitalized, they automatically decrease their physical activity by virtue of being in a room that centers around a bed and a television set. If they are hospitalized secondary to a physical symptom or illness, patients often feel the need to remain in bed because of the emotional trauma of their illness and the anticipation of extensive diagnostic testing. Additionally, the family encourages this presumed need for bed rest. Not only does the decrease in physical activity alter the final results of 24 hour pH testing but it also makes interpretation of the test more difficult when trying to establish a reflux pattern. If patients remain supine during the majority of the day while awake, they curtail one of the normal acid clearance mechanisms, i.e., gravity. Furthermore, if they happen to doze off while in the supine position, two other clearance mechanisms are diminished: saliva and esophageal contractions. Often the patient fails to record these seemingly brief naps, and acid exposure during this supine time is inaccurately reported as upright reflux.

The importance of ambulation was exemplified in a study by Schofield and associates,[20] who investigated 52 patients with angina pectoris and normal coronary angiograms. Eleven patients had abnormal baseline reflux patterns, nine of whom had their typical chest pain occur coincident with reflux during treadmill testing. More interestingly, 13 patients with normal baseline reflux values experienced their typical chest pain while on the treadmill, which coincided with acid reflux episodes. This latter group would have been misdiagnosed as having chest pain unrelated to gastroesophageal reflux if they had been studied only in the routine inpatient setting. In another study by Johnsson and coworkers,[10] the post-prandial period during the 24 hour pH study was compared with the combined ambulant and recumbent periods. These patients were encouraged to be as active as possible, at home or at work. Once again, reproducing the physiologic state as accurately as possible by encouraging ambulation increased the ability to discriminate between normal and abnormal reflux.

Two studies have confirmed that outpatient pH testing more accurately reflects pathologic acid reflux states. Schlesinger and colleagues[6] investigated the ability of 24 hour pH testing to discriminate inpatients with reflux symptoms (with and without erosive esophagitis) from inpatients without symptoms. Only 48 percent of patients with symptoms had abnormal pH studies. In those patients with erosive esophagitis, 29 percent had false-negative studies. These authors suggested that inpatient esophageal pH testing may underestimate the amount of acid reflux, and outpatient studies could be more precise. Branicki and co-workers[26] confirmed this hypothesis by studying 10 reflux patients in the hospital and at home or work using a pH-sensitive radiotelemetry capsule. Both the frequency and duration of acid reflux episodes were significantly greater in the out-patient compared with the inpatient setting. In fact, reflux parameters were usually five to ten times greater in the home and work environment. On the other hand, Jamieson et al.[22] generally found no significant differences among multiple reflux parameters when 10 reflux patients were studied in an inpatient setting and a second time in an unrestricted outpatient environment. The overall reproducibility of the various pH parameters exceeded 85 percent, but was 72 percent for the number of reflux episodes and percentage acid exposure time in the supine procedure. This was due to the patients having more reflux episodes in the ambulatory state and possibly sleeping better resulting in less nocturnal reflux.

There are occasions when esophageal pH testing would be more advantageously done on an inpatient basis. These instances include situations in which the patient's hospital stay would be too lengthy to wait for discharge or a diagnostic dilemma must be resolved before the patient goes home. A particular patient may also live too far from the medical center to return for further testing. In those locations where vandalization and/or theft of the pH monitor is of concern, the hospital setting would provide a more secure environment.

## SUMMARY

In conclusion, the preferable method for detecting abnormal amounts of esophageal acid exposure is ambulatory pH testing done in the outpatient setting over 24 hours. Although shorter studies have good specificity and reasonable sensitivity, only 24 hour pH studies consistently have excellent sensitivity and reproducibility. Furthermore, longer studies allow a better assessment of symptoms, particularly in patients with suspected atypical presentations of gastroesophageal reflux disease. Patients' tolerability and equipment cost are essentially equal for all time periods. Since ambulation and lifestyle habits affect acid reflux patterns, utilizing the outpatient setting during pH testing provides the most dependable approach for reproducing patients' usual physiologic environment.

## REFERENCES

1. Fink SM, McCallum RW: The role of prolonged esophageal pH monitoring in the diagnosis of gastroesophageal reflux. *JAMA* 252:1160–1164, 1984.
2. Rokkas T, Anggiansah A, Uzoechina E, et al: The role of shorter than 24-hour pH monitoring periods in the diagnosis of gastro-oesophageal reflux. *Scand J Gastroenterol* 21:614–620, 1986.
3. MacMahon M, Murray J, Hogan B, et al: 3 hour vs 24 hour intra-oesophageal pH monitoring. *Gut* 30:A1492, 1989.

4. Jorgensen F, Elsborg L, Hesse B.: The diagnostic value of computerized short-term oe-sophageal pH monitoring in suspected gastro-oesophageal reflux. *Scand J Gastroenterol* 23:363–368, 1988.

5. Klauser AG, Heinrich C, Schindibeck NE, Muller-Lissner SA: Is long term esophageal pH monitoring of clinical value? *Am J Gastroenterol* 84:362–365, 1989.

6. Schlesinger PK, Donahue PE, Schmid B, Layden TJ: Limitations of 24-hour intraesophageal pH monitoring in the hospital setting. *Gastroenterology* 89:797–804, 1985.

7. Porro GB, Pace F: Comparison of intraesophageal pH recording in the diagnosis of gastro-esophageal reflux. *Scand J Gastroenterol* 23:743–750, 1988.

8. Fuchs KH, DeMeester TR, Albertucci M: Specificity and sensitivity of objective diagnosis of gastroesophageal reflux disease. *Surgery* 102:575–580, 1987.

9. Vitale GC, Cheadle WG, Sadek S, et al: Computerized 24-hour ambulatory esophageal pH monitoring and esophagogastroduodenoscopy in the reflux patient. *Ann Surg* 200:724–729, 1984.

10. Johnsson F, Joelsson B, Isberg PE: Ambulatory 24 hour intraesophageal pH-monitoring in the diagnosis of gastroesophageal reflux disease. *Gut* 28:1145–1150, 1987.

11. Walther B, DeMeester TR: Comparison of 8-H and 16-H esophageal pH monitoring. In De-Meester TR, Skinner DB, eds: *Esophageal Disorders: Pathophysiology and Therapy*. New York, Raven Press, 1985:589–591.

12. Grande L, Pujol A, Ros E, et al: Intraesophageal pH monitoring after breakfast and lunch in gastroesophageal reflux. *J Clin Gastroenterol* 10:373–376, 1988.

13. Choiniere L, Miller L, Ives R, Cooper JD: Comparison of 8-hour studies after 24-hour studies. In: DeMeester TR, Skinner DB, eds: *Esophageal Disorders: Pathophysiology and Therapy*. New York, Raven Press, 1985:583–588.

14. Dobhan R, Castell DO. Prolonged intraesophageal pH monitoring with 16 hr overnight record-ing. Comparison with "24-hr" analysis. *Dig Dis Sci* 37:857–864, 1992.

15. Dhiman RK, Saraswat VA, Mishra A, Naik SR. Inclusion of supine period in short duration pH monitoring is essential in diagnosis of gastroesophageal reflux disease. *Dig Dis Sci* 41:764–72, 1996.

16. DeCaestecker JS, Heading RC: Esophageal pH monitoring. *Gastroenterol Clin North Am* 19:645–669, 1990.

17. Johnsson F, Joelsson B: Reproducibility of ambulatory oesophageal pH monitoring *Gut* 29:886–889, 1988.

18. DeCaestecker JS, Blackwell JN, Pryde A, Heading RL: Daytime gastro-oesophageal reflux is important in oesophagitis. *Gut* 28:519–526, 1987.

19. Sadek S, Cheadle W, Cranford C, et al: Patterns of gastroesophageal reflux associated with oe-sophagitis. *Gut* 25:A1143, 1985.

20. Schofield PM, Bennett DH, Whorwell PJ, et al: Exertional gastro-oesophageal reflux: A mecha-nism for symptoms in patients with angina pectoris and normal coronary angiograms. *Br Med J* 294:1459–1461, 1987.

21. Branicki FJ, Evans DF, Ogilvie AL, et al: Ambulatory monitoring of oesophageal pH in reflux oesophagitis using a portable radiotelemetry system. *Gut* 23:992–998, 1982.

22. Jamieson JR, Stein HJ, DeMeester TR, Bonavina L, Schwizer W, Hinder RA, Albertucci M: Ambulatory 24 hr esophageal pH monitoring: normal values, optimal thresholds, specificity, sensitivity, and reproducibility. *Am J Gastroenterol* 87:1102–1111, 1992.

# 6

# Influence of Diet, Smoking, Alcohol, and Exercise on Gastroesophageal Reflux: Possible Effects on pH Monitoring

Donald O. Castell, M.D.

Numerous reports have been published on the influence of dietary and other life-style factors in promoting gastroesophageal reflux in both normal individuals and patients with chronic reflux disease. Much of this information is concerned with effects on the lower esophageal sphincter (LES) and the potential for sustained or transient decreases in pressure, allowing reflux to occur. There are some more recent studies that have specifically evaluated distal esophageal acid exposure following a variety of commonly ingested foods or following specific activities. Whether these factors have an important effect on prolonged pH monitoring when used as a diagnostic modality is less clear, since little information is available to specifically address this point.

This chapter is divided into three parts: the first addresses the effects of diet, smoking, alcohol, and exercise on LES pressures; the second discusses more recent studies on the effect of these various factors on esophageal acid exposure; and the final section attempts to answer the question of the possible influence of important dietary or behavioral factors on diagnostic ambulatory pH monitoring.

# EFFECTS OF DIETARY AND LIFESTYLE FACTORS ON LOWER ESOPHAGEAL SPHINCTER PRESSURE

The earliest studies on the potential effects of foods, smoking, and alcohol on gastro-esophageal reflux were performed while monitoring LES pressures. It was postulated that a significant decrease in resting sphincter pressure would result in increased distal esophageal acid exposure secondary to the decreased competence of the antireflux barrier. Confirmation of actual increases in reflux was usually not obtained because of the limitations in the technology for accurate and convenient prolonged pH monitoring. In one of the earlier studies, Nebel and Castell monitored the effects of specific meals of protein, carbohydrate, and fat on LES pressure for 1 hour postprandially.[1] These studies in normal volunteers revealed that a high protein meal caused increases in LES pressures, whereas a carbohydrate meal (glucose solution) produced no significant change in sphincter pressure. The most striking observation from these early studies was the dramatic decrease in LES pressure produced by fat ingestion. For at least 1 hour following a corn oil meal, sustained significant decreases in sphincter pressure were recorded. These results are summarized in Figure 6.1. A similar effect was subsequently reported after ingestion of defatted chocolate syrup, with significant decreases in LES pressure occurring for approximately 1 hour after chocolate ingestion.[2] The reported effect of peppermint on LES pressure and reflux potential was quite different than that shown with fat and chocolate ingestion. In the studies by Sigmund and MacNally, transient decreases in LES pressure of only approximately 20 to 30 seconds' duration occurred within the first 10 minutes after ingestion of a solution of essence of peppermint by normal volunteers.[3] In similar experiments in normal volunteers, the acute administration of intoxicating amounts of ethanol by the oral or intravenous route was shown to produce decreases in LES pressure.[4] These observations have provided the basis for the current concepts of diet modification as a form of primary therapy for patients with reflux.

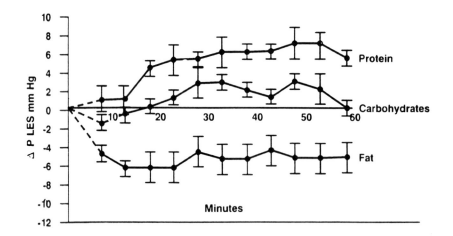

**Figure 6.1.** Changes in mean LES pressure from baseline for 60 minutes after ingestion of a high protein meal, a carbohydrate solution, and a fat meal. Vertical lines indicate ± standard error.

Cigarette smoking has been shown to produce significant decreases in LES pressure that persist throughout the duration of active smoking.[5] These effects were not sustained, however, with pressures returning to normal baseline values within a few seconds after cessation of smoking. Although the question is of considerable interest, there are no data available on the effect of exercise on LES pressures. The failure to obtain this information is related primarily to the technical difficulties of maintaining constant pressure recordings in a sphincter during the active movements occurring with exercise.

## EFFECTS OF DIETARY AND LIFESTYLE FACTORS ON DISTAL ESOPHAGEAL ACID EXPOSURE

The earlier studies cited above showed a variety of effects of foods, smoking, and alcohol on LES pressures and suggested that these factors might play a role in producing gastroesophageal reflux. However, the studies suffered from the absence of corroborating esophageal pH testing. With the development of improved technology and computer analysis for prolonged pH monitoring, it became possible to evaluate the actual effects of some of these factors on reflux. There has been good support in recent literature for the hypothesis that food-related effects on LES pressure may well translate into increased potential for gastroesophageal reflux. Studies in normal subjects have compared distal esophageal acid exposure for up to 3 hours after meals of low (16 percent) and high (61 percent) fat content. As summarized in Figure 6.2, these studies revealed that normal individuals have significantly increased amounts of distal esophageal acid exposure for up to 3 hours with the high fat meal when they are in the upright position post-prandially.[6] In contrast, there was increased esophageal acid exposure noted for 3 hours after the meal

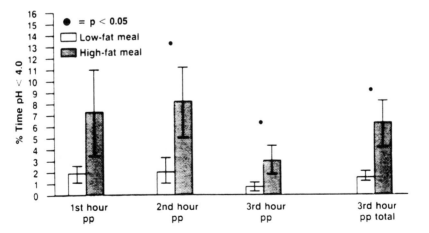

**Figure 6.2.** Mean percentage of time in the upright position with intraesophageal pH below 4.0 for normal volunteers following ingestion of a low fat (16 percent) and a high fat (61 percent) meal. Results are shown for each of the first 3 hours postprandially (pp) and for the total 3 hours after the meal. Vertical lines indicate ± standard error.

with the subjects in the supine position, but no differences in acid exposure could be found between the high and low fat meals. In patients with chronic reflux symptoms, the high fat meal did not promote any significant increase in distal esophageal acid exposure compared with the low fat meal in either the upright or the supine position postprandially. As with the normal subjects, however, significant increases in reflux were shown for up to 3 hours after both meals in the supine (recumbent) position, as shown in Figure 6.3. It was concluded that the high fat meal was much more likely to have an effect in normal individuals, in whom a lowering of LES pressure might decrease the competence of a normal antireflux barrier. In the patients with chronic reflux, this effect was less likely to occur because the barrier was already compromised. A recent study using intra-esophageal pH monitoring has confirmed the effect of high fat and also large volume of meals to induce reflux.[7]

Similar to the studies with fatty meals, the effect of defatted chocolate syrup on esophageal acid exposure has also been studied for 1 hour after the chocolate ingestion. These studies demonstrated a significant increase in esophageal acid exposure during the first hour.[8]

A recent study has provided some confirmation that peppermint can increase reflux. Compounds of the carminative class, containing volatile fatty acids, include peppermint, onions and garlic. In a recent study it was shown that the addition of onions to a hamburger test meal resulted in significant increases in distal esophageal acid exposure for up to 2 hours after the meal in patients with chronic heartburn.[9] Similarly, a recent study using prolonged pH monitoring has confirmed the potential for alcohol to increase reflux. In healthy volunteers, 4 ounces of Scotch whiskey (40% alcohol) ingested in the evening produced significant increases in nocturnal acid exposure compared with those of a control study.[10]

Another study has also confirmed that cigarette smoking acutely increases acid reflux events in both normal smokers and smokers with gastroesophageal reflux disease.[11] It was proposed that the mechanisms of acid reflux during cigarette smoking were mainly dependent on the presence of a chronically low LES pressure. In contrast, there were signif-

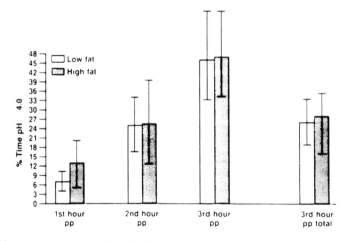

**Figure 6.3.** Average percentage of time in the supine position for intraesophageal pH below 4.0 in patients with gastroesophageal reflux following ingestion of a low fat (16 percent) and a high fat (61 percent) meal. Results are shown for each of the first 3 hours postprandially (pp) and for the total 3-hour postprandial interval. Vertical lines indicate ± standard error.

icant increases in *transient* LES relaxations during smoking, although the majority of reflux episodes occurred with coughing or deep inspiration; during which increases in intra-abdominal pressure might overcome a compromised sphincter (i.e., "stress reflux").

A report of the effects of exercise on reflux studied by intraesophageal pH monitoring has shown differences that were dependent upon the type of exercise.[12] Vigorous exercise on both a stationary bicycle and through a series of Nautilus weight exercises produced only minor increases in gastroesophageal reflux. In contrast, during a 15-minute period of running, significantly increased reflux was seen, as shown in Figure 6.4. When these normal volunteers were further tested for a full 60 minutes of running, significant reflux was noted throughout the hour.[13] The reflux episodes were frequent and usually brief, and most were accompanied by belching. It was proposed that the more active form of exercise associated with increased bodily movement such as running may readily allow gastric acid to reflux, particularly when the LES transiently relaxes to relieve air swallowed while exercising.

Recently, recumbent position has been shown to effect reflux following a meal. Greater acid exposure was seen in normal volunteers when lying with the right side down compared to left side down.[14]

It is commonly believed that patients who are obese are more likely to suffer from increased amounts of gastroesophageal reflux. In fact, many regimens for the treatment of patients with reflux symptoms include weight loss as a component of the baseline therapy often termed lifestyle modifications. Yet, there is little evidence in the literature to even suggest that body weight, per se, changes the tendency for reflux independent of the effect of dietary fat. Nor are there any clinical studies indicating that weight loss changes reflux parameters in patients with abnormal acid exposure who happen to be obese.[15]

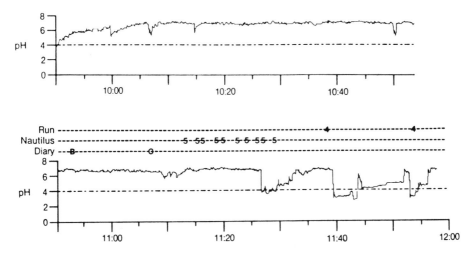

**Figure 6.4.** Two hours of intraesophageal pH recording for one normal volunteer during a basal hour (above) and during 1 hour of various exercises (below). A pH scale is shown on the vertical axis and actual time on the horizontal axis. The 15 minute bicycling interval is indicated by the letters on the line marked "Diary," the 15 minute series of Nautilus exercises are indicated by the numbers "5" on the appropriate line, and the 15 minute run is included between the numbers "4" on the appropriate line. Reflux is indicated when the patient's pH recording falls below the level of 4.0 as indicated by the broken line.

# EFFECTS OF DIETARY AND LIFESTYLE FACTORS ON pH TESTING

The studies discussed above provide ample evidence that foods such as fat, chocolate, and carminatives (onions, peppermint) may have a significant effect on esophageal acid exposure. Likewise, alcohol, smoking, and exercise can also increase, distal esophageal acid exposure, at least transiently. What is not clear, however, is how important these factors might be during diagnostic testing for reflux with ambulatory prolonged pH monitoring. In one study attempting to evaluate this question relative to smoking, it was concluded that smoking did not adversely affect the results of 24 hour pH monitoring.[16] More episodes of reflux were found in smokers than nonsmokers, particularly in the upright position, and a trend toward greater total reflux in smokers. There were, however, no significant differences in the total time with pH exposure less than 4.0 during 24 hour pH testing.

Many patients with chronic reflux note their symptoms predominantly during ingestion of large fatty meals or after chocolate and alcohol and some report symptoms are precipitated by exercise. To obtain an accurate test of their usual reflux pattern, therefore, it becomes important to specifically include, rather than exclude, these activities during a 24 hour sample of their lifestyle. The patient should maintain their diary to allow accurate assessment of those factors that might be contributing to reflux but these should not be avoided if an accurate representation of their reflux activity is to be obtained. Finally, it is important to exclude the actual time of meal ingestion from the overall analysis of reflux episodes and percent time of acid exposure.[17] As shown in Figure 6.5, frequent decreases

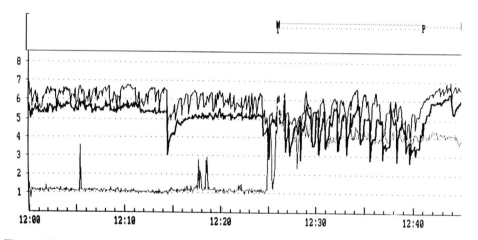

**Figure 6.5.** Forty-five minutes of intraluminal pH recording during a 24 hour study in a patient with chronic reflux symptoms. The lower recording shows intragastric pH and the two upper recordings distal (heavy line) and proximal esophageal pH. The time during meal ingestion (indicated by the M and dotted horizontal line) has been ignored (indicated by | and dotted line) from the analysis because of the frequent decreases of esophageal pH to < 4 produced by slightly acidic material traveling *down* the esophagus which would otherwise be inappropriately interpreted as acid reflux traveling *up* the esophagus.

of intraesophageal pH below 4.0 may occur during eating, particularly related to the ingestion of coffee, tea, cola, tomato products, or citrus juices. For this reason, we specifically exclude the period of meal ingestion during the overall analysis of reflux.*

# REFERENCES

1. Nebel OT, Castell DO: Lower esophageal sphincter pressure changes after food ingestion. *Gastroenterology* 63:778–783, 1972.
2. Wright LE, Castell DO: The adverse effect of chocolate on lower esophageal sphincter pressure. *Am J Dig Dis* 20:703–709, 1975.
3. Sigmund CJ, McNally EF: The action of a carminative on the lower esophageal sphincter. *Gastroenterology* 56:13–18, 1969.
4. Hogan WJ, Viegas de Andrade SR, Winship DH: Ethanol induced acute esophageal motor function *J Appl Physiol* 32:755–760, 1972.
5. Dennish GW, Castell DO: The inhibitory effect of smoking on the lower esophageal sphincter. *N Engl J Med* 284:1136, 1137, 1971.
6. Becker DJ, Sinclair J, Castell DO, et al: A comparison of high and low fat meals on postprandial esophageal acid exposure. *Am J Gastroenterol* 84:782–786, 1989.
7. Iwakisi K, Kobayashi M, Kotoyori M, et al: Relationship between postprandial esophageal acid exposure and meal volume and fat content. *Dig Dis Sci* 41:926–930, 1996.
8. Murphy DW, Castell DO: Chocolate and heartburn: Evidence of increased esophageal acid exposure after chocolate ingestion. *Am J Gastroenterol* 83:633–636, 1988.
9. Allen ML, Mellow MH, Robinson MG, et al: The effect of raw onions on acid reflux and reflux symptoms. *Am J Gasdtroenterol* 84:377–380, 1990.
10. Vitale GC, Cheadle WG, Patel B, et al: The effect of alcohol on nocturnal gastroesophageal reflux. *JAMA* 258:2077–2079, 1987.
11. Kahrilas PJ, Gupta RR: Mechanisms of acid reflux associated with cigarette smoking. *Gut* 31:4–10, 1990.
12. Clark CS, Kraus BB, Sinclair JW, et al: Gastroesophageal reflux induced by exercise in healthy volunteers. *JAMA* 261:3599–3601, 1989.
13. Kraus BB, Sinclair JW, Castell DO: Gastroesophageal reflux in runners. Characteristics and treatment. *Ann Intern Med* 112:429–433, 1990.
14. Katz LC, Just R, Castell DO: Body position affects recumbent postprandial reflux. *J Clin Gastroenterol* 18:280–283, 1994.
15. Castell DO: Obesity and gastroesophageal reflux: Is there a relationship? *European Journal Gastroenterol Hepatol* 8:625, 626, 1996.
16. Schindlbeck NE, Heinrich C, Dendafer A, et al: Influence of smoking and esophageal intubation on esophageal pH-metry. *Gastroenterology* 92:1994–1997, 1987.
17. Wo JM, Castell DO: Exclusion of meal periods from ambulatory pH monitoring may improve diagnosis of esophageal acid reflux. *Dig Dis Sci* 39:1601–1607, 1994.

*Editor's note:* This routine exclusion of the meal time is not usually done in our laboratory or most others because the meals per se effect the total and upright acid exposure times minimally. However, excessively long meals or frequent ingestion of acidic beverages may cause a false positive study and need to be excluded (see Chapter 3, Figure 3.6).

# 7

# Methods of Data Analysis

## Lawrence F. Johnson, M.D.

Although in the early 1970s continuous intraesophageal pH monitoring provided a much needed solution for measuring gastroesophageal reflux objectively, it unfortunately presented a complex problem. The problem consisted of how best to define or score that objectivity. The complexity of the problem is apparent when the different pH values that could serve as criteria for acid reflux and its esophageal clearance are considered. Moreover, these different criteria could be counted as absolute number, percent, or frequency of occurrence. In addition, these criteria could be tabulated according to time segments (day or night) or other phenomena that occurred during a period of continuous intraesophageal pH monitoring, such as posture (upright or recumbent) and mental state (awake or asleep). Finally, there are different methods of data analysis to score the pH record on the basis of the above criteria. Hence, the effective use of this new diagnostic technology required that order be created from chaos.

This chapter addresses the logic we[1] and others used at the time to select certain pH values and criteria to measure reflux and esophageal acid clearance objectively. The reader is reminded that our six criteria and the 24 hour composite score were published over 22 years ago. Thus, this chapter discusses why it was done then and whether it is relevant now. The reader will be given a better background with which to judge the scientific merit of both the old and new literature on different methods of data analysis for continuous esophageal pH monitoring.

## DEVELOPMENT OF 24 HOUR pH COMPOSITE SCORE

A distal drop in intraesophageal pH to a value below 4.0 was defined as reflux of acid gastric contents because it was an established reference point during other pH reflux tests and had been shown to denote the onset of heartburn. In retrospect, this was a good choice, because Wallin and Madsen[2] have subsequently shown that a pH value of less than 4.0 as opposed to less than 3.0 or less than 5.0, gave the best discrimination for establishing normal values for a group of asymptomatic volunteers. Moreover, Vitale and colleagues[3] showed that a pH value of less than 4.0 discriminated best between symptomatic patients and controls.

We placed the pH probe 5 cm above the proximal margin of the LES, as determined by esophageal manometry. This site was chosen because we wanted an esophageal location that had precedence for phenomena related to gastroesophageal reflux. Precedent for this location came from multiple observations. First, this location was identical to that used for the standard acid reflux test[4] and seemed to be in close proximity to that correlating heartburn with an intraluminal drop in pH (i.e., 5 cm above the respiratory inversion point of the LES).[5] Second, this location appeared to be in the immediate area that earlier investigators had used in positioning their probes by fluoroscopy, that is, 5 cm above the cardia,[6] diaphragmatic hiatus,[7] or the junction of the middle and lower one third of the esophagus.[8] Our use of esophageal manometry to consistently place the pH probe 5 cm above the LES may have eliminated some variability in the data, especially since it has recently been shown that the degree of esophageal acid exposure in normal individuals decreases significantly in the proximal esophagus.[9,10] Even though placement of the pH probe 5 cm above the LES has served as a standard location, the question can still be asked: Is this the best location? This author feels that the question has probably not been studied adequately.

Although others had described continuous intraesophageal pH monitoring techniques consisting of 6 to 18 hours,[8,11] we felt that patient tolerance would permit extending the time period to 24 hours. Although it was easy to discern that the circadian cycle should be divided into a day and nighttime segment, the problem of how to divide and define the segments was not clear. For instance, when did day end and night begin, i.e., using the sunset as influenced by seasonal change or using the electric switch as determined by hospital policy or patient preference? Alternatively, the circadian cycle could be divided into an awake versus a sleep period, but to expect medical personnel to make this judgment or to define sleep polygraphically was impractical. We settled ultimately on posture to divide and define the two segments of the circadian cycle. While awake, the individual would remain in the upright posture (sitting or standing) and at the appropriate times eat breakfast, lunch, and supper, as determined by hospital policy. At night, bedtime would be determined by individual preference, and the patient would sleep in the recumbent posture. Hence, posture defined both segments of the circadian cycle so that its influence on the reflux pattern could be evaluated separately.

Defining protocol policy for circadian cycle time segments for ambulatory outpatient intraesophageal pH monitoring is no less important today than it was 23 years ago for in hospital monitoring studies. In fact, a recent study[12] examining the hourly pattern of distal esophageal acid exposure showed that although patients exceeded controls, both groups had more exposure during the evening hours (1700 to 2400 hours) than during the day or nighttime segment. If confirmed, this observation could affect the dose and schedule of administration for various therapeutic agents.

The six parameters used in the 24 hour pH composite score were selected on the basis of (1) our understanding of the pathophysiology of gastroesophageal reflux disease in 1973, (2) clinical observations obtained while patients were being monitored, and (3) the experience of other authors. At the time it was commonly accepted that acid exposure and reflux events were the generic markers of the disease. This led us to use acid exposure expressed as percentage of time the pH was below 4.0[8] and its measurement during the upright and recumbent periods as well as for the total 24 hour period. This selection afforded measurement of acid exposure not only for two different segments of the circadian cycle but also for the total 24 hour period. For the same reasons, the total number of reflux episodes for the entire 24 hour period was also tabulated.

Although the concept of esophageal acid clearance had been embodied in a test,[13] it was clinically presumed that reflux events during the day were similar to those at night. Instead, we found that reflux events in symptomatic patients during the upright period occurred frequently and, in general, were rapidly cleared. In contrast, those that occurred at night were less frequent and, in general, required longer clearing periods. These varying clearance times prompted us to measure the time required for the pH to return to 4.0 or more following each reflux event (i.e., the esophageal acid clearance time). For parameters that concerned esophageal acid clearance, we selected the number of reflux episodes lasting more than or equal to 5 minutes and the single longest reflux episode. Precedents existed for these two measurements because Miller and Doberneck[14] had shown that control subjects seldom had reflux episodes lasting 5 minutes or longer. In addition, they had used a single 15 minute reflux episode in a symptomatic patient as an indication for antireflux surgery. Although we considered tabulating our two esophageal acid clearance parameters for both the upright and recumbent segments of the circadian cycle, we feel that a single 24 hour tabulation would suffice.

Having chosen the above six parameters by studying symptomatic patients with gastroesophageal reflux disease, we did not know the normal values for these parameters. To establish these values, we monitored asymptomatic control volunteers and found a characteristic reflux pattern. This pattern consisted of rapidly cleared reflux episodes immediately after meals, occasionally between meals, and only rarely at night. We termed this "physiologic reflux." Abnormality for each of the six parameters was identified as any value that exceeded the mean for the control group by more than two standard deviations. This definition was chosen because statistically 95 percent of the asymptomatic population would have a value within this limit if the values for each parameter fell into a normal distribution pattern.

When applying the above definition of abnormality to a patient population with typical gastroesophageal reflux symptoms, the six parameters were not uniformly sensitive (Table 7.1). For instance, the parameter with the highest incidence of abnormality was acid exposure during the recumbent period. In contrast, the total number of reflux episodes for the 24 hour period had the lowest incidence of abnormality. These data showed that patients with symptoms of gastroesophageal reflux did not have a homogeneous reflux pattern, thereby emphasizing the need for multiple parameters to collectively define abnormal degrees of gastroesophageal reflux.

Although the six parameters provided a means for identifying any conceivable abnormal reflux pattern, it remained unclear what constituted patient abnormality, i.e., deviation from one, two, three, or all six parameters. Although we thought initially about assigning points to an abnormal parameter, thereby developing a scoring system, the magnitude of points assigned to any given parameter was arbitrarily determined by us instead of letting the data from the asymptomatic controls determine the scoring unit. To solve this problem, we decided to use the standard deviation (SD) found for each parameter observed in the asymptomatic control volunteer group as the scoring unit for that parameter. To award points, the scoring unit or standard deviation was divided into the pa-

**TABLE 7.1. DERIVATION OF 24 HOUR pH SCORING***

| Parameters Mean ± 1 SD | Asymptomatic Control (N = 15) | Normal Value† | Symptomatic Patients with Abnormal Values (%)† (N = 38) |
|---|---|---|---|
| Recumbent period | 0.286% ± 0.467 | <1.2% | 82 |
| Total period | 1.478% ± 1.381 | <<4.2% | 78 |
| No. of episodes ≥5 | 0.6 ± 1.241 | ≤3 | 74 |
| Longest episode (minutes) | 3.866 ± 2.689 | <9.2 | 74 |
| Upright period | 2.33% ± 1.975 | <6.3% | 63 |
| Total of episodes | 20.6 ± 14.773 | <50 | 50 |

From Johnson LF, DeMeester TR: Twenty-four pH monitoring of the distal esophagus: A quantitative measure of gastro-esophageal reflux. *Am J Gastroenterol* 62:325–332, 1974.

*A value within the mean and two standard deviations for the asymptomatic controls.

†$P \leq 0.01$ for all six parameters between the symptomatic patients and the asymptomatic controls.

tient's absolute value for that parameter. This method sensitized the awarding of points in accordance with the ability of that parameter to define abnormality. For instance, normal individuals have a moderate number and wide spectrum of values for the parameter total number of reflux episodes per 24 hours (i.e., large intersubject variability). Thus, this results in a high mean value and a large standard deviation. For this parameter, the standard deviation serving as a scoring unit reflects the wide spectrum of values found in normal individuals and in turn awards fewer points for that parameter when it is observed in symptomatic patients. In contrast, normal individuals rarely experience reflux at night; therefore, the mean value for nighttime acid exposure is low, as is the standard deviation (i.e., small intersubject variability). Thus, the scoring unit awards more points for nocturnal acid exposure, which is a marked departure from physiologic reflux. Hence, using the standard deviation as a scoring unit allows the data from normal individuals to score the pH record of symptomatic patients uniformly in a manner weighted to their departure from physiologic reflux. In addition, this scoring unit deals equitably with each of the six parameters, even though they may be different (acid exposure, reflux episodes, esophageal acid clearance phenomena) all obtained during different segments of the circadian cycle.

In order to use the standard deviation as the scoring unit, a zero point had to be established for each parameter. This was accomplished by designating zero as two standard deviations below the actual mean value for that particular parameter (Figure 7.1). Two standard deviations were necessary, since only one designated a point either slightly above or slightly below the actual value of zero for a given parameter. Hence, two standard deviations below the actual mean value of that parameter established a new or theoretical zero point to which a value from any patient could be compared and in turn points awarded by the scoring system. By establishing a new zero point in this manner for each of the six parameters, scoring would be equitable among the parameters, even though their individual mean values and standard deviations varied.

The patient's 24 hour pH composite score was determined by adding all the points assessed for each parameter. An abnormal 24 hour pH composite score was defined as one

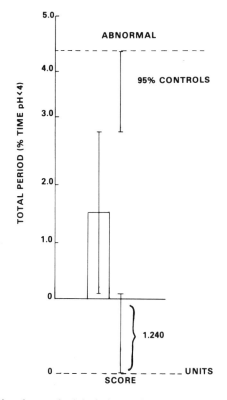

**Figure 7.1.** Concept of using the standard deviation as the scoring unit. Note the establishment of a zero score two standard deviations below the mean value for total period of acid exposure. (From Johnson LF, DeMeester TR: Twenty-four hour pH monitoring of the distal esophagus: A quantitative measure of gastroesophageal reflux. *Am J Gastroenterol* 62:325–332, 1974.)

that exceeded the mean and two standard deviations attained for the 15 asymptomatic control volunteers (i.e., greater than or equal to 22). It is pertinent to note that all asymptomatic volunteers had a score within two standard deviations of the mean value determined for their group, and 84 percent of patients (N = 38) with typical symptoms of gastroesophageal reflux had an abnormal score.[1] In later studies, the 24 hour composite score gave comparable results in additional patients and controls (90 to 94 percent sensitivity and 90 to 100 percent specificity).[15,16] To facilitate scoring the pH record, tables were later established that represented a continuum of advancing degrees of normality and abnormality for potential values of the six parameters and their respective points that would be awarded by the scoring unit for that parameter (Table 7.2). For instance, acid exposure (percentage of time pH is less than 4.0) during the supine period for the patient shown in Table 7.2 is 2.0 percent. When this absolute value (2 percent) is related to the table, 4.67 points are awarded and placed in the column labeled Score. In a similar manner, points are derived from the table for the absolute values of the other five reflux parameters and placed in the score column adjacent to their respective parameters. The 24 hour pH composite score is then derived by totaling all the points in the score col-

**TABLE 7.2. TABULATION OF 24 HOUR INTRAESOPHAGEAL pH COMPOSITE SCORE**

| Parameters | Points Awarded | Score |
|---|---|---|
| Supine period (% time pH <4.0) | | |
| 0.25 | 0.92 | |
| — | — | |
| 2.0* | 4.67 | 4.67 |
| 2.25 | 5.20 | |
| 2.50 | 5.74 | |
| — | — | |
| 40.0 | 86.04 | |
| Total period | | — |
| Number of reflux episodes ≥5 minutes | — | |
| Longest reflux episode (minutes) | | — |
| Upright period | | — |
| Total number of reflux episodes | | — |
| 24-HOUR COMPOSITE SCORE = | | TOTAL POINTS |

*Example patient value.

umn. Alternatively, the 24 hour composite score could be determined by a computer program which is commercially available and known as the Johnson-DeMeester Composite Score.

Although the 24 hour pH composite score was developed in 1973, its concept is still consistent with advances in our understanding of the nature of gastroesophageal reflux. For instance, reflux was formerly thought to be a pathologic event, but the scoring system recognized and allowed for physiologic reflux. Moreover, the scoring system harshly scored any reflux pattern that deviated from that of the physiologic pattern found in normal individuals. Although gastroesophageal reflux was not even thought of as a nocturnal disease in 1973, the scoring system recognized nocturnal acid exposure as a marked departure from physiologic reflux and scored exposure during this segment of the circadian cycle highly. This recognition was important because in subsequent studies, we[17,18] and others[16,19] have shown that large degrees of nocturnal acid exposure (percentage of time pH is less than 4.0) resulting from frequent and/or poorly cleared reflux events represent a risk factor for esophagitis, strictures, and Barrett's esophagus.

Although it was tempting to include only nocturnal-related parameters in the 24 hour pH composite score, some patients have abnormal acid exposure only during the daytime segment of the circadian cycle. It is clinically important to identify these upright refluxers, since they do not respond well to a fundoplication because of the gas bloat syndrome, but they may respond to unorthodox therapy such as biofeedback.[20] For these patients, the 24 hour pH composite score also has identifying parameters such as acid exposure in the upright position and total number of reflux episodes. The fact that their abnormal score tends to be low denotes their diminished propensity to develop reflux esophagitis. Hence, the scoring system is clinically relevant to reflux patterns found during both segments of the circadian cycle.

# OTHER SCORING METHODS

The record from 24 hour intraesophageal pH monitoring can be scored by four other concepts. No single concept, including the 24 hour pH composite score,[1] is universally accepted. A scoring system that uses only parameters with high reproducibility has diagnostic merit, especially since a single study conducted over 24 hours (i.e., 1/360 of a year) will determine whether a symptomatic patient has a normal or abnormal degree of gastroesophageal reflux. Wiener and coworkers[21] examined multiple pH parameters to determine which had the highest concordance rates for intrastudy variability. They found that acid exposure for the total, supine, and upright periods had among the three highest concordance values (85 percent, 80 percent, and 75 percent, respectively) for the 12 parameters examined. In their study, the 24 hour pH record was considered abnormal if any value from one or more of the above three parameters exceeded the mean and two standard deviations found in their asymptomatic control population. Using this scoring system, 93 percent of the patients with reflux esophagitis (N = 14) and the asymptomatic controls (N = 14) remained in the proper group after two 24 hour intraesophageal pH monitoring studies performed within 10 days of each other.

The second concept requires defining a limited time segment (i.e., several hours) from the record that best correlates with results from the entire 24 hour pH record. Patients are monitored only during this abbreviated period, and the record is scored. Two recent studies attempted to address this issue. In both studies[22,23] the authors performed 24 hour intraesophageal pH monitoring in asymptomatic individuals and patients with clinical and/or endoscopic evidence of reflux esophagitis. Time segments were looked for that afforded the same sensitivity and specificity in classifying the study population as that of the entire 24 hour record. Although both studies did not agree on the best short time interval to replace the 24 hour period, it seems reasonable to further investigate the use of the postprandial period after breakfast and lunch for an abbreviated test in the event that the physician does not want to monitor the patient during sleep.

Third, the 24 hour record can be scored by making a product of the reflux events multiplied by their duration.[3,24,25] For instance, when a product of the total number of reflux events that occurred during a 15 hour overnight monitoring period (6 P.M. to 10 A.M.) was multiplied by their mean duration (minutes), the resulting product or score afforded a 90 percent sensitivity for the study population.[24] However, in another study of longer duration (24 hours), the mean number of reflux episodes per hour was multiplied by the mean cumulative duration of reflux per hour (minutes), giving a product or score with only a 76 percent sensitivity.[3] This lower sensitivity could have resulted from using a single mean value for each parameter (i.e., frequency and duration) derived by combining the hourly values obtained during two diametrically opposed central nervous system states (awake versus sleep) and postures (upright versus recumbent) as well as for disproportionate periods of time (16 hours daytime and 8 hour nighttime). Thus, in order to effectively score pH records with a reflux event × duration product, it appears prudent to count reflux events that occur during a predominant time segment that represents a marked departure from physiologic reflux (i.e., the nocturnal period). Moreover, the reflux events that occur should be at risk for delayed esophageal acid clearance. Last, one should not determine a mean for hourly values that occur during diametrically opposed central nervous system states or postures, and for disproportionate periods of time.

The fourth concept for scoring the pH record concerns establishing a symptom index score. This score elucidates whether acid reflux episodes recorded during pH monitoring relate directly to symptoms experienced concurrently by the patient.[26] The symptom index score was defined as the number of times a symptom such as heartburn was reported by the patient while the pH was less than 4.0 divided by the total times the symptom was noted. This fraction was then multiplied by 100, giving the result as a percentage. Thus, a symptom index score could be obtained for almost any symptom and thereby address the clinical question: Did the acid reflux episodes recorded by pH monitoring relate to the patient's symptoms? The authors showed a good correlation between a high symptom index score for heartburn (75 to 100 percent) and an abnormal pH record (97.5 percent) as well as between a low symptom index score (0 to 24 percent) and a normal record (78 percent). In addition, the authors showed that a high or low symptom index score for heartburn related appropriately to either the presence or the absence of reflux esophagitis. Thus, the symptom index scoring system differs in concept from other systems that use asymptomatic controls to establish cut-off values for abnormal reflux parameters, thereby only inferring that patients' symptoms are related to pathologic degrees of gastroesophageal reflux. Most important, the symptom index score does not preclude the use of any other scoring system, since it is independent of "physiologic or pathologic" degrees of gastroesophageal reflux.

Regardless of the scoring system used for interpreting the record from any ambulatory outpatient intraesophageal pH monitoring study, asymptomatic control volunteers still need to be initially monitored. The study of controls is necessary, especially since the interrelationship of diet, timing of meals, physical activity level, sleep, and posture influences physiologic reflux. For instance, naps taken during the day or meals eaten late at night followed by recumbency and sleep affect the degree of physiologic reflux. Thus, in setting up a new protocol, it is important to remember that the degree of physiologic reflux that occurs in controls establishes normal values that later serve as the standard against which symptomatic patients are judged. To establish cut-off values that best designate physiologic versus pathologic reflux may warrant the use of a receiver operating curve to analyze data from both controls and patients. This statistical concept would appear indicated, especially since pH monitoring data from controls does not have a normal distribution.[27] Most important, patients studied during any new ambulatory pH monitoring protocol should not be compared with controls from other studies who may not have been monitored under comparable circumstances.

# REPRODUCIBILITY

A high degree of reproducibility has been shown for various parameters of intraesophageal pH monitoring.[21,28] Reflux parameters that measure acid exposure (percentage of time pH is less than 4.0) have the highest correlation coefficients, and those that measure the total number of reflux events per unit time have the lowest correlation coefficients. Of those parameters that measured acid exposure, the total 24 hour period seemed to give the highest concordance values (85 and 77 percent), followed by the upright (75 and 83 percent) and supine (80 and 62 percent) periods. Although a poor concordance value could be attributed to the monitoring technique or a poor time segment chosen for analysis, it must be remembered that the patient and/or disease factors may affect reproducibility. For instance, there is the potential for a great deal of variability during two up-right periods owing to size and timing of meals, physical activity levels, posture, naps, bowel movements, and smoking habits that affect reflux and could contribute to a low concordance rate. In contrast, during the supine period, sleep and arousal from sleep profoundly affect the esophageal acid clearance time.[29] Thus, it is conceivable that differences in the sleep pattern between the two nights could account for a low concordance rate.

Alternatively, it could be asked why a reflux parameter with a high concordance value such as total acid exposure does not have perfect reproducibility (a correlation coefficient that approaches 1.0). Interestingly, the 24 hour intraesophageal pH monitoring technique may be at fault. For example, when two pH probes are placed in the esophagus at the same level, they do not give an identical tracing.[30] This study raises concern that on occasion the pH electrode may abut directly against the esophageal mucosa and not give an accurate reading of intraluminal pH. Regardless of the cause, the most variability between adjacent pH probes seems to occur at night during the supine period, suggesting that some reflux events may be missed, especially those with long clearance times.

Intrasubject reproducibility supports the diagnostic use of continuous intraesophageal pH monitoring because the variability in values does not appreciably affect the proper grouping of patients and controls. However, the percentage of change in values for some parameters would restrict its use for measuring a response to some therapeutic agents.[21] For example, to statistically prove that a small decrease in esophageal acid exposure (total percentage of time pH is less than 4.0) resulted from a proposed therapeutic agent would require a prohibitive number of patients (Table 7.3). Thus, only those therapies that are anticipated to render major reductions in acid exposure (greater than or equal to 33 percent) should be subjected to documentation by continuous intraesophageal pH monitoring.[21]

**TABLE 7.3. ESTIMATED PATIENT SAMPLE SIZES FOR PLACEBO-CONTROLLED CLINICAL TRIALS AFFECTING TOTAL PERCENT TIME pH <4.0***

| Percent Reduction in Total % Time pH <4.0 Due to Drug | Required Sample Size in Each Group† | |
| --- | --- | --- |
| | 80% Power | 90% Power |
| 10 | 402 | 557 |
| 20 | 90 | 125 |
| 25 | 54 | 75 |
| 33 | 28 | 40 |
| 40 | 18 | 25 |
| 50 | 10 | 14 |

*From Wiener GJ, Morgan TM, Copper JB, et al: Ambulatory 24-hour esophageal pH monitoring. Reproducibility and variability of pH parameters. *Dig Dis Sci* 33:1127–1133, 1988.

†For comparing a single post-treatment minus pretreatment percentage time pH <4.0 between the treatment groups at the 0.05 one-sided level of significance with the stated power.

# FORECAST FOR CONTINUOUS
# INTRAESOPHAGEAL pH MONITORING

It is only natural that as intraesophageal pH monitoring evolves, modifications will occur in the criteria for counting acid reflux events and their esophageal clearance as well as in the methods for data analysis. For instance, although a drop in intraesophageal pH to less than 4.0 has strong merit for the diagnosis of acid reflux, it is not a perfect criterion. A recent study comparing pH and scintigraphic evidence of gastroesophageal reflux simultaneously showed that pH monitoring missed reflux events during the early postprandial period owing to neutralization of gastric contents by the meal.[31] Moreover, after an acid reflux event had occurred and the intraluminal pH was less than 4.0 additional reflux events occurred but were not counted because of the conventional definition of a reflux event. This failure to count reflux events that occur during the acid clearing process could be addressed by modifying the current definition of a pH reflux event. The modification would state that when the intraesophageal pH is less than 4.0 and drops 2 or more units, this drop and any other(s) of similar degree denote additional reflux event(s). This proposed modification was supported by scintigraphic evidence of reflux.[31]

A recent French study[32] proposed an alternative to counting the number of acid reflux events and measuring their clearance time from the esophagus. This new concept consists of measuring the area under the curve for pH values of less than 4.0 on the 24 hour pH record. This approach not only addresses acid exposure analogous to the parameter percentage of time pH is less than 4.0 but also factors in the magnitude of drop in acid intraesophageal pH. This concept would also put in perspective the annoying phenomena whereby intraesophageal pH oscillates just above and below a value of 4.0, such that these events and their clearance require tabulating even though they are probably not of clinical relevance. Thus, as continuous intraesophageal pH monitoring evolves it will undergo changes that should improve its ability to accurately diagnose acid reflux and its esophageal clearance.

# SUMMARY

There are many different concepts for analyzing data from the 24 hour intraesophageal pH monitoring record. Implemented concepts should incorporate into their design the proper relationship of circadian events such as alimentation, posture, and sleep. Although data from some reflux parameters can be pooled for the 24 hour period (total percentage of time pH is less than 4.0), other parameters, such as the reflux frequency and esophageal acid clearance time, are affected by different physiologic events that occur within the circadian cycle. For these parameters, it is not advisable to determine one mean value for the entire 24 hour period. Since the phenomenon of gastroesophageal reflux is not pathologic, cut-off values that best designate abnormality need to be established by studying large groups of asymptomatic controls and representative symptomatic patients. Judging whether a given patient's reflux values are pathologic can only be accomplished by a direct comparison with the cut-off values obtained from a study population monitored in a similar environment under an identical protocol. Moreover, even when pathologic degrees of gastroesophageal reflux are verified, the patient's symptoms during the monitoring period still need to be correlated with the reflux events in order to ensure that these events caused the symptoms.

Although 24 hour intraesophageal pH monitoring established the concept of physiologic reflux, provided insights into the mechanisms causing esophagitis, and has served as a diagnostic reference standard for pathologic acid gastroesophageal reflux, it is not a perfect test. Both the technique itself and patient factors related to physiologic events may contribute to the discordance observed between two tests when conducted in the same patient within a short period of time. Improving the criteria that denote acid reflux as well as defining limitations concerning the technique of 24 hour intraesophageal pH monitoring can be accomplished by other tests that measure gastroesophageal reflux. The development of new equipment for 24 hour ambulatory outpatient esophageal pH monitoring may have reached its zenith; however, improvements in reflux criteria and new innovations for data analysis should remain active. A thorough knowledge of the investigators' thought processes that pioneered this test along with how their principles have held up over time should aid the reader in judging the merit of new concepts and methods of data analysis for 24 hour intraesophageal pH monitoring.

# REFERENCES

1. Johnson LF, DeMeester TR: Twenty-four hour pH monitoring of the distal esophagus: A quantitative measure of gastro-esophageal reflux. *Am J Gastroenterol* 62:325–332, 1974.
2. Wallin L, Madsen T: Twelve-hour simultaneous registration of acid reflux and peristaltic activity in the oesophagus: A study in normal subjects. *Scand J Gastroenterol* 14:561–566, 1979.
3. Vitale GC, Cheadle WG, Sarvi S, et al: Computerized 24-hour ambulatory esophageal pH monitoring and esophagogastro-duodenoscopy in the reflux patient: A comparative study. *Ann Surg* 200:724–729, 1984.
4. Kantrowitz PA, Carson JG, Fleischli DG, et al: Measurement of gastroesophageal reflux. *Gastroenterology* 56:666–674, 1969.
5. Tuttle SG, Rufin F, Battaneloo A: The physiology of heartburn. *Ann Intern Med* 55:292–300, 1961.
6. Woodward DAK: Response of the gullet to gastric reflux in patients with hiatal hernia and oesophagitis. *Thorax* 24:459–464, 1970.
7. Pattrick FG: Investigation of gastroesophageal reflux in various positions with a two-lumen pH electrode. *Gut* 11:659–667, 1970.
8. Spencer J: Prolonged pH recording in the study of gastroesophageal reflux. *Br J Surg* 56: 912–914, 1969.
9. Johansson KE, Tibbling L: Evaluation of the 24-hour pH test at two different levels of the esophagus. In: DeMeester TR, Skinner DB, eds: *Esophageal Disorders: Pathophysiology and Therapy.* New York: Raven Press, 1985:579–582.
10. Sondheimer JM, Haase GM: Simultaneous pH recordings from multiple esophageal sites in children with and without distal gastroesophageal reflux. *J Pediatr Gastroenterol Nutr* 7:46–51, 1988.
11. Miller FA: Utilization of inlying pH probe for evaluation of acid peptic diathesis. *Arch Surg* 89:199–203, 1964.
12. Gudmundsson K, Johnsson F, Joelsson B: The time pattern of gastroesophageal reflux. *Scand J Gastroenterol* 23:75–79, 1988.

13. Booth DG, Kemmerer WI, Skinner DB: Acid clearing from the distal esophagus. *Arch Surg* 96:731–734, 1968.

14. Miller FA, Doberneck RC: Diagnosis of the acid-peptic diathesis by continuous pH analysis. *Surg Clin North Am* 47:1325–1334, 1967.

15. DeMeester TR, Wang CL, Wernly JA, et al: Technique, indications and clinical use of 24 hour esophageal pH monitoring. *J Thorac Cardiovasc Surg* 79:656–670, 1980.

16. Pujol A, Grande L, Ros E, et al: Utility of inpatient 24-hour intraesophageal pH monitoring in diagnosis of gastroesophageal reflux. *Dig Dis Sci* 33:1134–1140, 1988.

17. Shay SS, Eggli D, McDonald C, et al: Gastric emptying of solid food in patients with gastro-esophageal reflux. *Gastroenterology* 92:459–465, 1987.

18. Orr WC, Lackey C, Robinson MG, et al: Esophageal acid clearance during sleep in patients with Barrett's esophagus. *Dig Dis Sci* 33:654–659, 1988.

19. Robertson D, Aldersley M, Shepherd H, et al: Patterns of acid reflux in complicated oesophagitis. *Gut* 28:1484–1488, 1987.

20. Shay SS, Johnson LF, Wong RKH, et al: Rumination, heartburn and daytime gastroesophageal reflux: A case study with mechanisms defined and successfully treated with biofeedback therapy. *J Clin Gastroenterol* 8:115–126, 1986.

21. Wiener GJ, Morgan TM, Copper JB, et al: Ambulatory 24-hour esophageal pH monitoring. Reproducibility and variability of pH parameters. *Dig Dis Sci* 33:1127–1133, 1988.

22. Porro GB, Pace F: Comparison of three methods of intra-esophageal pH recordings in the diagnosis of gastroesophageal reflux. *Scand J Gastroenterol* 23:743–750, 1988.

23. Grande L, Pujol A, Ros E, et al: Intraesophageal pH monitoring after breakfast and lunch in gastroesophageal reflux. *J Clin Gastroenterol* 10:373–376, 1988.

24. Stanciu S, Hoare RC, Bennett JR: Correlation between manometric and pH tests for gastro-oesophageal reflux. *Gut* 18:536–540, 1977.

25. Branicki FJ, Evans DF, Jones JA, et al: A frequency duration index (FDI) for the evaluation of ambulatory recordings of gastroesophageal reflux. *Br J Surg* 71:425–430, 1984.

26. Wiener GJ, Richter JE, Copper JB, et al: The symptom index: A clinically important parameter of ambulatory 24-hour esophageal pH monitoring. *Am J Gastroenterol* 83:358–361, 1988.

27. Schindlbeck NE, Heinrich C, Konig A, et al. Optimal thresholds, sensitivity, and specificity of long-term pH-metry for the detection of gastroesophageal reflux disease. *Gastroenterology* 93:85–90, 1987.

28. Johnsson F, Joelsson B: Reproducibility of ambulatory oesophageal pH monitoring. *Gut* 29:886–889, 1988.

29. Orr WC, Johnson LF, Robinson MG: The effect of sleep on swallowing, esophageal peristalsis, acid sensitivity and clearance time. *Gastroenterology* 86:814–819, 1984.

30. Murphy DW, Yuan Y, Castell DO: Does the intraesophageal pH probe accurately detect acid reflux? *Dig Dis Sci* 34:649–656, 1989.

31. Shay SS, Eggli D, Johnson LF: Simultaneous esophageal pH monitoring and scintigraphy during the postprandial period in patients with severe reflux esophagitis. *Dig Dis Sci* 36:558–64, 1991.

32. Izquierdo MA, Tovar JA, Eizaguirre I: Esophageal acid exposure in a single number: The area under the pH curve. *Chir Pediatr* 30:1–5, 1989.

# 8

# Ambulatory 24 Hour Esophageal pH Monitoring—What Is Abnormal?

**Ross M. Bremner, M.D.**
**Tom R. DeMeester, M.D.**
**Hubert J. Stein, M.D.**

Since Miller's original description of the use of an indwelling pH probe to evaluate "acid-peptic diathesis,"[1] prolonged pH monitoring has been shown to be an accurate method for detecting changes in the pH environment of the esophagus and measuring esophageal exposure to gastric juice.[2,3] Following the availability of a portable digital data recorder in the last years, 24 hour esophageal pH monitoring has rapidly increased in its popularity and is currently the accepted means to quantitate reflux of gastric contents into the esophagus over a complete circadian cycle and to confirm the presence of gastroesophageal reflux disease. The information obtained from 24 hour esophageal pH monitoring provided an opportunity to conceptualize the pathophysiology of this complicated disease process and stimulated a rational, stepwise approach to determining the cause of abnormal esophageal exposure to gastric juice and to the design of specific therapy for correcting the abnormality.[4–7] As the technique has become more widely available, questions have been raised regarding the units to express esophageal exposure to gastric juice, the upper limits of normal esophageal acid and alkaline exposure, and the sensitivity and specificity of the measurement to detect gastroesophageal reflux disease.[8–12]

## HOW TO EXPRESS ESOPHAGEAL EXPOSURE TO GASTRIC JUICE

Esophageal exposure to gastric contents as measured by pH monitoring is usually expressed by the cumulative time the pH is outside the normal range during a 24 hour period. The nature of the exposure, i.e., acid or alkaline, is determined by measuring this time using different pH thresholds. Figure 8.1 shows the range of esophageal pH in 50

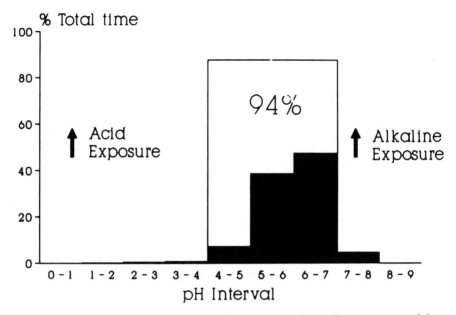

**Figure 8.1.** The normal range of esophageal pH expressed as the median percentage of the total time spent at each pH interval in 50 normal volunteers. Note that 94 percent of the time is spent within pH 4.0 to 7.0.

normal asymptomatic volunteers expressed as the median time spent at the pH intervals 0–1, 1–2, 2–3, 7–8, and 8–9 and illustrates that in the normal situation the pH is between 4.0 and 7.0 over 94 percent of the monitored time. In normal subjects, esophageal exposure to a pH outside this range is minimal. This indicates that esophageal exposure to acid or alkaline gastric contents can be measured by the percentage of time the pH is below 4.0 or above 7.0, respectively.

Technical factors can artificially increase the measured esophageal acid or alkaline exposure. These include the use of antimony probes, the drift of an unstable electrode, or errors in calibration.[13] We recommend using glass electrodes in all studies rather than antimony electrodes,* which can be unstable in alkaline or very acidic pH ranges, and to check electrodes in buffered solutions before *and* after each test to assure proper calibration and the absence of drift.

Abnormally high esophageal exposure to pH greater than 7.0 is a less dependable indicator of gastroesophageal reflux than anatomically high exposure to a pH less than 4.0.[14] An increase in the percentage of time the pH is above 7.0 in the distal esophagus can also be attributable to the presence of dental infection, which increases the salivary pH, ingestion of food with a greater than 7.0 pH, or to the presence of esophageal obstruction, which may result in static pools of saliva with bacterial overgrowth and a rise in esophageal luminal pH. To avoid these errors, special precautions should be taken. Pa-

*Editor's note:* A consensus panel in 1987 suggested that "nearly all devices on the market irrespective of the type of electrode used" are adequate for clinical studies assessing acid reflux disease.[13]

tients' diets during the test should be restricted to a list of foods with a pH between 4.0 and 7.0. Each patient should have an assessment of his or her oral hygiene, and all esophageal strictures must be dilated prior to pH monitoring to prevent pooling of saliva. Only assiduous attention to these factors allows reliable measurement of acid *and* alkaline gastroesophageal reflux with 24 hour pH monitoring.

The percent time the esophagus is exposed to a pH of greater than 7 has historically been used as an indirect indicator of the reflux of duodeno-gastric juice. Studies have shown that the esophageal exposure time to a pH > 7 was increased in patients who had Barrett's esophagus with complications such as stricture, ulceration and dysplasia versus those without complications (Figure 8.2).[15] An important observation is that the mean value was different between the two groups but both were within the normal range as defined by the 95th percentile of normals. From the perspective of prevalence data, the importance of an increased exposure to pH > 7 is clearly seen in patients with complications (Figure 8.3). A more accurate means of detecting esophageal exposure to duodenal juice is now available with a spectrophotometric device which measures the wavelength of bilirubin (see Chapter 17). Studies are now emerging using the Bilitec 2000 that support the previous findings of increased esophageal exposure to duodenal contents in patients with complicated Barrett's disease.[16-19]

The single measurement of time the pH is below or above a threshold, although concise, does not reflect how the exposure occurred, e.g., did it occur in a few long episodes

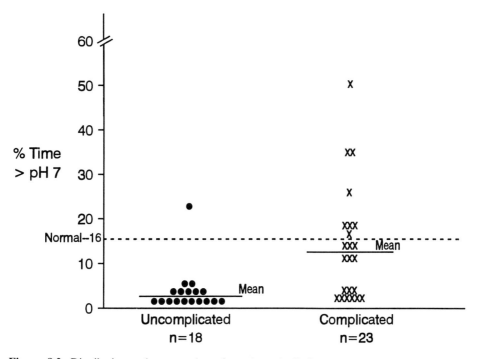

**Figure 8.2.** Distribution and mean value of esophageal alkaline exposure (percentage of time pH > 7) in patients with or without complications of Barrett's esophagus ($p < 0.01$).

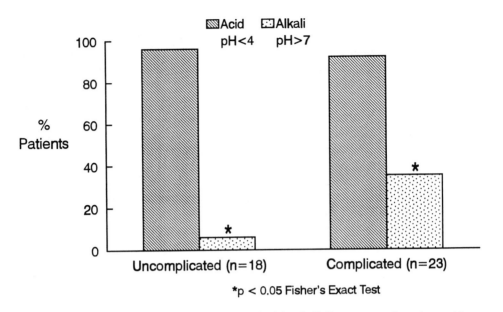

**Figure 8.3.** The prevalence of abnormal esophageal acid and alkaline exposure in patients with or without complications of Barrett's esophagus.

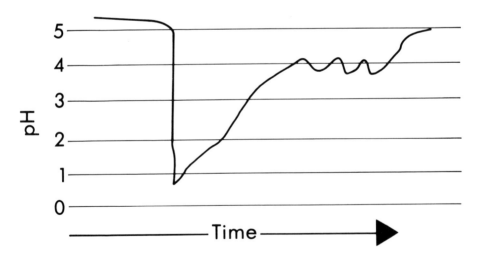

**Figure 8.4.** Illustration of the difficulty in measuring the duration of a reflux episode. The problem is in determining the completion of an episode, i.e., when the pH rises above 4.0 for the first time or when it stays above 4.0 for a set period of time. The most precise definition is the former and allows for data evaluation without judgment. (From DeMeester TR, Stein HJ: Gastroesophageal reflux disease. In: Moody FG, Carey LC, Jones RC, et al., eds: *Surgical Treatment of Digestive Disease*, ed 2. Chicago: Year Book Medical Publishers, 1989:68, with permission.)

or several short episodes? Consequently, it is necessary to measure the frequency of each exposure below or above a given threshold and its duration. To measure these requires defining when a reflux episode has ended. A problem in this regard emerges when the esophageal pH at the end of a reflux episode fluctuates just above and below the given threshold before rising or dropping definitely above or below the threshold.[13] Figure 8.4 illustrates the problem and shows how an acid reflux episode could be recorded as a single episode or as one long episode followed by three short episodes. We believe that a reflux episode is completed when the pH rises or drops above or below the threshold for the first time. This provides the most precise means of measuring frequency and duration and allows for collection of data without judgment.

A 24 hour period is generally felt to be an adequate duration for pH monitoring in that it allows the measurement of esophageal exposure to gastric juice over one complete human circadian cycle. This period allows the monitoring of the effect of physiologic activity, such as eating in the upright position and sleeping in the recumbent position. In order to obtain comparable data for the meal and the upright and supine periods, we instruct our patients to remain in the upright or sitting position while awake and to ingest three meals at the appropriate times. The meals are unique only in the absence of food and beverages with a pH value less than 4.0 or greater than 7.0. The times of meals and sleep are recorded by having the patient keep a diary in which food and fluid intake, the time the supine position is assumed in preparation for sleep, and the time or rising in the morning are noted. Hence, the influence of both segments of the circadian cycle on esophageal exposure to gastric juice can be evaluated separately, that is, while the patient is awake, in the upright posture before, during, and after meals, and while he or she is sleeping in the recumbent posture at night.

Based on experience in patients and volunteers, we chose to express the results of the 24 hour pH recording in the following ways: (1) cumulative reflux time, expressed as the percentage of total monitored time, percentage of time monitored with the patient upright, and percentage of time monitored with the patient supine; (2) frequency of reflux episodes, expressed as the number of reflux episodes per 24 hours; and (3) duration of reflux episodes, expressed as the number of episodes longer than 5 minutes and the number of minutes of the longest episode.[6,20] Not all of these six parameters are uniformly abnormal in patients with gastroesophageal reflux disease. For instance, the component most commonly abnormal is acid exposure during the recumbent period, whereas the total number of reflux episodes per 24 hours has the lowest incidence of abnormality. In order to combine the six components of the 24 hour esophageal pH record into one holistic value, we developed a composite score that weighs each component according to its dependability and reliability and allows determination of a normal or abnormal pH record for each pH threshold.[3,6,20] The development and details of this scoring system are described in Chapter 7.

# WHAT IS ABNORMAL ESOPHAGEAL
# EXPOSURE TO GASTRIC JUICE?

To identify pathologic gastroesophageal reflux, it is necessary to define how much esophageal exposure to gastric juice is normal or abnormal. The lack of a gold standard in the diagnosis of gastroesophageal reflux disease made us conclude that it would be easier to define normality than it would to define the presence of the disease. Therefore, we measured esophageal exposure to gastric juice by 24 hour esophageal pH monitoring in 50 normal subjects (20 males, 30 females) who were free of symptoms or objective evidence of any foregut abnormality on physical examination, esophageal manometry, and a barium swallow.[21] This approach is analogous to defining dwarfism by first determining normal height distribution rather than finding and measuring those who appear to be dwarfs.

Twenty-four hour esophageal pH monitoring was performed in the 50 normal subjects using a combined glass electrode with a built-in reference electrode placed 5 cm above the upper border of the manometrically determined lower esophageal sphincter. The electrodes were calibrated at pHs 1.0 and 7.0 at the beginning and end of the monitoring period, and all studies with an electrode drift greater than 0.2 pH units were excluded. Food and beverages during the study were limited to those with a pH value between 4.0 and 7.0. Subjects were fully ambulatory during the monitoring period and instructed to perform normal daily activities. The pH data were acquired at a frequency of 0.25 Hz and stored on a portable digital data recorder. To measure the pH exposure outside the normal pH range of 4.0 to 7.0, the time spent below the pH thresholds 4.0, 3.0, 2.0, and 1.0 and above the thresholds 7.0 and 8.0 was calculated and expressed as the percentage of time the pH was below or above these thresholds for the total monitored time and the time spent in the upright and supine positions. To define the nature of this exposure, the total number of episodes, the number of episodes longer than 5 minutes, and the time of the longest episode in minutes were calculated for each pH threshold.

Since measurement of esophageal pH values is not normally distributed, percentiles are required to establish a normal range.[13] The mean, median, and 95th percentile for esophageal acid exposure (pH less than 4.0) in the normal volunteers for each of the six components are given in Table 8.1. The medians and 95th percentiles for all pH thresholds can also be graphically displayed, as shown in Figures 8.5 A through F. This allows the comparison of a patient's results for each component with a large data base of normal volunteers and the identification of an abnormal exposure with a confidence of 95 percent. The normal values for each of the six components when compared among three major centers in the United States and various centers throughout the world show an unexpected uniformity, indicating that esophageal exposure to gastric juice in normal individuals is similar despite nationality or dietary habits.[13,22] Consequently, symptomatic patients whose degree of esophageal exposure to gastric juice exceeds that of the normal subjects are considered to have gastroesophageal reflux disease, provided the measure-

TABLE 8.1. VALUES OF 24 HOUR ESOPHAGEAL pH MONITORING IN 50 HEALTHY VOLUNTEERS FOR pH <4.0

| | Mean | Standard Deviation | Median | Min. | Max. | 95th Percentile |
|---|---|---|---|---|---|---|
| % Total time pH < 4.0 | 1.5 | 1.4 | 1.2 | 0 | 6.0 | 4.5 |
| % Upright time pH < 4.0 | 2.2 | 2.3 | 1.6 | 0 | 9.3 | 8.4 |
| % Supine time pH < 4.0 | 0.6 | 1.0 | 0.1 | 0 | 4.0 | 3.5 |
| No. of episodes | 19.0 | 12.8 | 16.0 | 2.0 | 56.0 | 46.9 |
| No. of episodes ≥ 5 minutes | 0.8 | 1.2 | 0 | 0 | 5.0 | 3.5 |
| Longest episode (minutes) | 6.7 | 7.9 | 4.0 | 0 | 46.0 | 19.8 |
| Composite score | 6.0 | 4.4 | 5.0 | 0.4 | 18.0 | 14.7 |

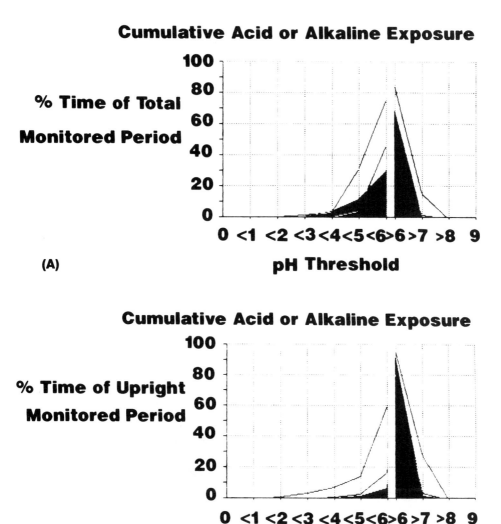

(A)

(B)

**Figure 8.5.** Graphic display of the six components of esophageal pH exposure showing the median and 95th percentile levels in 50 normal volunteers using whole pH values above and below 6.0 as thresholds. The black area represents measurements made in a patient. When the black area exceeds the 95th percentile line for a given pH threshold, the patient is considered to have an abnormal value for the component measured. (A) percent cumulative exposure for total time; (B) percent cumulative exposure for upright time.

*Continued*

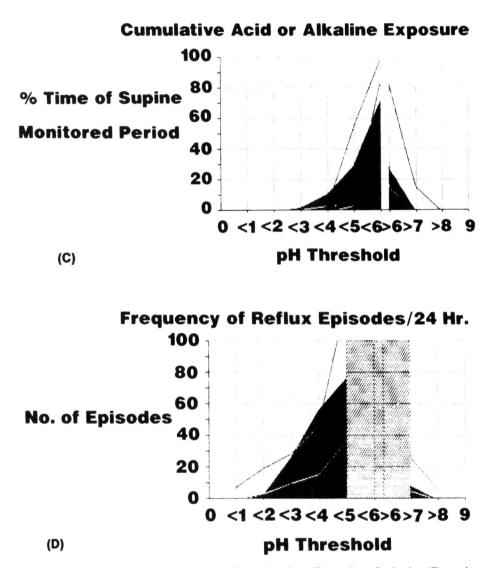

**Figure 8.5.** (C) percent cumulative exposure for supine time; (D) number of episodes; (E) number of episodes lasting longer than 5 minutes; (F) length of longest reflux episode. (From DeMeester TR, Stein HJ: Gastroesophageal reflux disease. In: Moody FG, Carey LC, Jones RC, et al., eds: *Surgical Treatment of Digestive Disease*, ed 2. Chicago: Year Book Medical Publishers, 1989:70–71, with permission.)

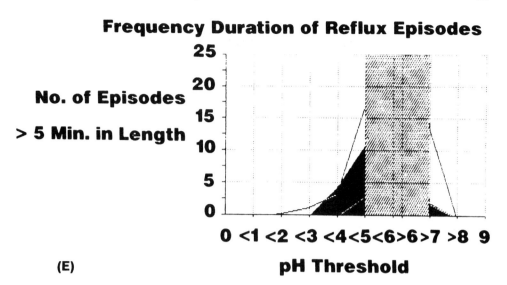

**Frequency Duration of Reflux Episodes**

**No. of Episodes**

**> 5 Min. in Length**

(E)

**pH Threshold**

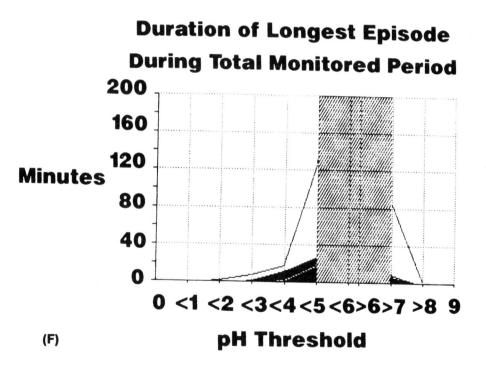

**Duration of Longest Episode**

**During Total Monitored Period**

**Minutes**

(F)

**pH Threshold**

**TABLE 8.2. UPPER LIMIT OF NORMAL (95TH PERCENTILE) OF THE COMPOSITE SCORE FOR VARIOUS pH THRESHOLDS**

| pH Threshold | Upper Limit of Normal (95th Percentile) |
|---|---|
| pH < 1.0 | 14.20 |
| pH < 2.0 | 17.37 |
| pH < 3.0 | 14.10 |
| pH < 4.0 | 14.72 |
| pH < 5.0 | 15.76 |
| pH < 6.0 | 12.76 |
| pH > 7.0 | 14.90 |
| pH > 8.0 | 8.50 |

ment has been performed according to the guidelines outlined above. Typical symptoms, or complications of the disease, such as esophagitis, stricture, or Barrett's esophagus, do not have to be present for the diagnosis to be made.

If the six components are combined into the composite score (see Chapter 7), an upper limit for a normal score, based on the 50 normal volunteers, can be constructed for each pH threshold (Table 8.2) and graphically displayed. A patient's value can then be superimposed on this graph and his or her esophageal pH exposure evaluated in relation to that measured in the normal subjects (Figure 8.6). An IBM-compatible program to perform this function is available from Gastrosoft, Inc. (1425 Greenway Drive, Suite 600, Irving, TX 75038).

When data on esophageal acid exposure in normal subjects from three major centers in the United States were combined, a large data base composed of information on 110 normal asymptomatic volunteers was created.[22] This data base allowed evaluation of the effects of age and gender on the normal values for the six components used to express esophageal acid exposure. The analysis showed that age did not affect the normal values for esophageal exposure to pH less than 4.0 except for an increased number of reflux episodes lasting longer than 5 minutes observed in subjects older than 50 years of age. Men, however, were found to have more physiologic reflux of acid gastric juice in that they had an increased percentage of time the pH was less than 4.0 during the total and upright time, an increased duration of the longest reflux episode, an increase in the number of reflux episodes, and an increase in the number of episodes lasting longer than 5 minutes. There is, however, no difference between males and females when the composite score is used to express esophageal acid exposure.[21] This would indicate that in patients with a borderline high exposure to a pH less than 4.0, gender has to be taken into account when interpreting the 24 hour pH record. The composite score for esophageal acid exposure is applicable to both males and females.

A comparison of inpatient and ambulatory outpatient monitoring in both normal subjects and patients showed no significant difference in the parameters used to measure esophageal acid exposure, with the exception that the number of reflux episodes experienced during the outpatient monitoring period were significantly increased in normal subjects.[21] This is not surprising, since an increase in the number of reflux episodes would be expected when a person is ambulatory. In patients with an abnormal test result, the environment had little influence on the number of reflux episodes.*

*Editor's note:* See Chapter 5 for a slightly different viewpoint.

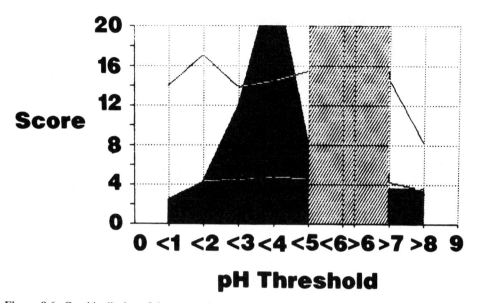

**Figure 8.6.** Graphic display of the composite score used to express the overall result of a 24 hour esophageal pH recording. The lower line represents the median score and the upper line the 95th percentile of 50 normal subjects. The black area represents the composite score of the patient shown in Figure 8.3 with increased esophageal acid exposure measured at pH less than 4.0. (From DeMeester TR, Stein HJ: Gastroesophageal reflux disease. In: Moody FG, Carey LC, Jones RC, et al., eds: *Surgical Treatment of Digestive Disease*, ed 2. Chicago: Year Book Medical Publishers, 1989:72, with permission.)

# SENSITIVITY AND SPECIFICITY OF 24 HOUR ESOPHAGEAL pH MONITORING IN THE DIAGNOSIS OF GASTROESOPHAGEAL REFLUX DISEASE

Knowledge of the sensitivity (the ability of the test to detect disease when it is known to be present) and the specificity (the ability of the test to exclude disease when it is known to be absent) is essential for the physician in order to use the test clinically. The sensitivity and specificity of 24 hour esophageal pH monitoring in the diagnosis of gastro-esophageal reflux disease in several large published series range between 79 and 95 percent and 86 and 100 percent, respectively, and are dependent on the threshold for ab-normal exposure and the criteria used to define gastroesophageal reflux disease. As shown in Figure 8.7 sensitivity and specificity of esophageal pH monitoring are higher than for any other objective test used in the diagnosis of the disease.

The construction of receiver operating characteristic curves illustrates the continuous relationship between the sensitivity and specificity of a test over a whole range of thresh-old values. The usefulness of receiver operating characteristic curves is to identify the threshold value of a test that yields optimal sensitivity and specificity. Using receiving op-erator characteristic analysis, we recently showed that the 95th and 96th percentiles of the

*Text Continues on page 92*

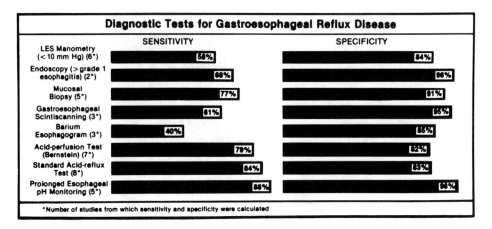

**Figure 8.7.** Sensitivity and specificity of diagnostic tests for gastroesophageal reflux disease in the recent literature. L.E.S. = lower esophageal sphincter. (From Bonavina L, DeMeester TR: Prolonged esophageal pH monitoring. In: Sigel B ed: *Diagnostic Patient Studies in Surgery.* Philadelphia, Lea & Febiger, 1986:358, with permission.)

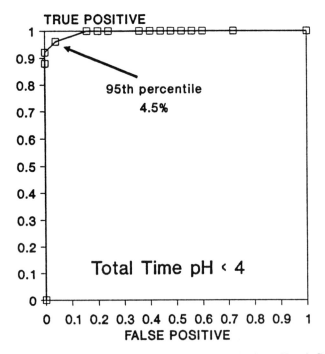

**Figure 8.8.** ROC curves for: A: % total time pH < 4, B: % upright time pH < 4, C: % supine time pH < 4.

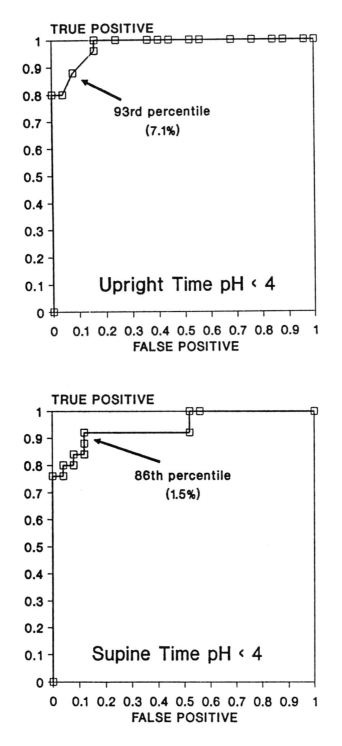

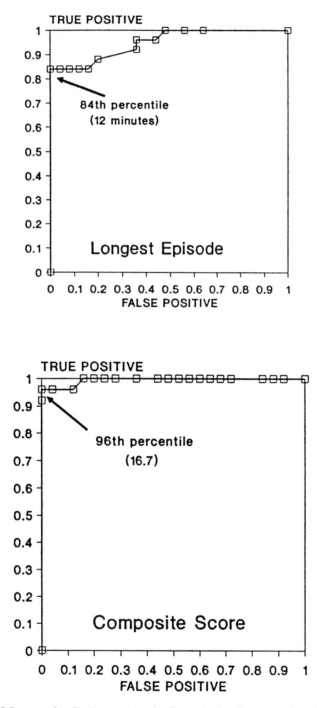

**Figure 8.8.** ROC curves for: D: the number of reflux episodes, E: the number of reflux episodes lasting longer than 5 minutes, F: the duration of the largest reflux episode, and G: the composite score. The point of maximum likelihood, i.e., the threshold that yields the optimal combination of sensitivity and specificity for each curve, is indicated by an arrow, and the percentile and absolute values are given.

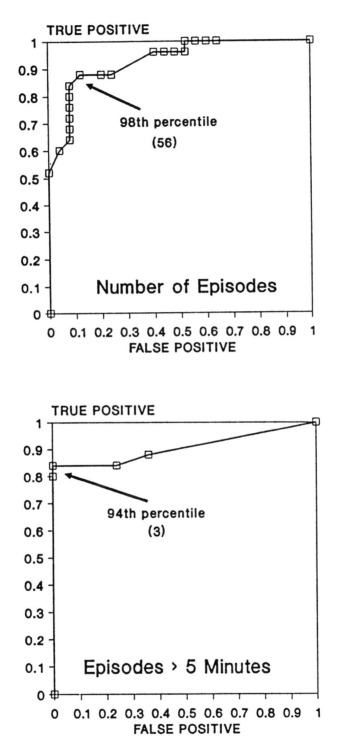

**TABLE 8.3. RECEIVER OPERATING CHARACTERISTICS FOR THE INTEGRATED SCORE AND ITS COMPONENT**

| Component | Optimal Percentile Threshold | Numerical Value |
|---|---|---|
| % Total Time pH < 4 | 95 | 4.5% |
| % Upright Time pH < 4 | 93 | 7.1% |
| % Supine Time pH < 4 | 86 | 1.5% |
| # Episodes | 98 | 56 |
| # Episodes > 5 Minutes | 94 | 3 |
| Longest Episode | 84 | 12.0 |
| Composite Score | 96 | 16.7 |

data obtained in our 50 normal volunteers for percentage of total time the pH is less than 4.0 and the composite score, respectively, provide the most efficient interpretation of 24 hour esophageal pH monitoring of all components used to measure exposure. (Figure 8.8A-B and Table 8.3) The sensitivity, specificity, positive predictive value, and accuracy of these parameters is shown in Table 8.4. On the basis of these studies, have the 95th percentile value for the percentage of total time the pH is less than 4.0 and the 96th percentile for the integrated score are the best determinants for detecting increased esophageal acid exposure. The values for these thresholds are 4.5 percent for the percentage of total time the pH is less than 4.0 and 16.7 for the composite score.

**TABLE 8.4. SENSITIVITY, SPECIFICITY, AND PREDICTIVE VALUE OF ESOPHAGEAL PH MONITORING IN THE DIAGNOSIS OF GASTROESOPHAGEAL REFLUX DISEASE DETERMINED AT THE POINT OF MAXIMUM LIKELIHOOD**

| | Sensitivity (%) | Specificity (%) | Positive Predictive Value (%) | Negative Predictive Value (%) | Accuracy (%) |
|---|---|---|---|---|---|
| % Total Time pH < 4 | 96 | 96 | 96 | 96 | 96 |
| % Upright Time pH < 4 | 92 | 88 | 92 | 88 | 90 |
| % Spine Time pH < 4 | 88 | 92 | 88 | 92 | 90 |
| # Episodes | 88 | 88 | 88 | 88 | 88 |
| # Episodes ≥ 5 Minutes | 84 | 100 | 100 | 86 | 92 |
| Longest Episode (minutes) | 84 | 100 | 100 | 86 | 92 |
| Composite Score | 96 | 100 | 100 | 96 | 98 |

# RELEVANCE OF SYMPTOMS DURING THE MONITORING PERIOD

Figure 8.9 shows that normal subjects experience the symptom of belching and coughing during the pH monitoring. These symptoms become less evident during the nighttime hours. Consequently it must be remembered when evaluating symptoms during the monitoring period that coughing and belching can be due to the presence of the pH probe and have less diagnostic value. In contrast heartburn and regurgitation are rarely reported by normal subjects during the monitoring period.

Clinical experience has shown that symptoms of chronic cough or wheezing can occur immediately before, during, after, or independently of a reflux episode, and the mechanical effects of courghing or sheezing can induce a reflux episode. Coughing and wheezing during or within 3 minutes of a reflux episode are probably reflex-induced due to the stimulation of esophageal receptors, whereas choking, spasms of cough, and shortness of breath are more apt to be due to actual aspiration. The dependability of the diagnosis of aspiration increases as progressively higher pH probes, i.e., upper esophagus, hypopharynx or trachea, record a drop in pH synchronous with symptoms. An important observation is that episodes of aspiration can be asymptomatic and can occur in patients who

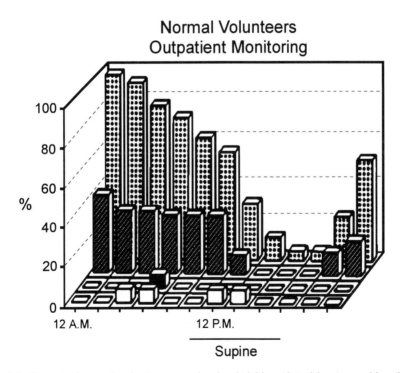

**Figure 8.9.** Percent of normal volunteers experiencing belching (dotted bars), coughing (hatched bars), heartburn (black bars), or regurgitation (white bars), during the 24-hour ambulatory outpatient monitoring.

have normal 24-hour esophageal acid exposure. Respiratory symptoms independent of reflux episodes are probably due to primary pulmonary disease or to bronchial mucosal damage from aspiration episodes too infrequent to be documented during a single 24-hour period of esophageal pH monitoring.

# REFERENCES

1. Miller FA: Utilization of inlying pH probe for evaluation of acid-peptic diathesis. *Arch Surg* 89:199–203, 1964.
2. DeMeester TR, Johnson LF: The evaluation of objective measurements of gastroesophageal reflux and their contribution to patient management. *Surg Clin North Am* 56:39–53, 1976.
3. Johnson LF, DeMeester TR: Twenty-four hour pH monitoring of the distal esophagus: A quantitative measure of gastroesophageal reflux. *Am J Gastroenterol* 62:325–332, 1974.
4. DeMeester TR, Wang CI, Wernly JA, et al: Technique, indications, and clinical use of 24-hour esophageal pH monitoring. *J Thorac Cardiovasc Surg* 79:656–670, 1980.
5. DeMeester TR, Johnson LF, Joseph GJ, et al: Patterns of gastroesophageal reflux in health and disease. *Ann. Surg* 184:459–470, 1976.
6. DeMeester TR, Stein HJ: Gastroesophageal reflux disease. In: Moody FG, Carey LC, Jones RC, et al, eds: *Surgical Treatment of Digestive Disease*, ed 2. Chicago: Year Book Medical Publishers, 1989:65.
7. DeMeester TR, Bonavina L, Albertucci M: Nissen fundoplication for gastroesophageal reflux disease—evaluation of primary repair in 100 consecutive patients. *Ann Surg* 204:9–20, 1986.
8. Wiener GJ, Morgan TM, Copper JB, et al: Ambulatory 24-hour esophageal pH monitoring. Reproducibility and variability of pH parameters. *Dig Dis Sci* 33:1127–1133, 1988.
9. Schlesinger, PK, Donahue, PE, Schmid B, et al: Limitations of 24-hour intraesophageal pH monitoring in the hospital setting. *Gastroenterology* 89, 797–804, 1985.
10. Johnsson F, Joelsson B: Reproducibility of ambulatory oesophageal pH monitoring. *Gut* 29:886–889, 1988.
11. Schindlbeck NE, Heinrich C, Köuig A, et al: Optimal thresholds, sensitivity and specificity of long-term pH-metry for the detection of gastroesophageal reflux disease. *Gastroenterology* 93:85–90, 1987.
12. Klausen AG, Heinrich C, Schindlebeck NE: Is long-term esophageal pH monitoring of clinical value? *Am J Gastroenterology*, 84:362–366, 1989.
13. Emde C, Garner A, Blum AL: Technical aspects of intraluminal pH-metry in man: Current status and recommendations. *Gut* 28:1177–1188, 1987.
14. Stein HJ, Barlow AP, DeMeester TR, et al: The development of complications in gastro-esophageal sphincter, esophageal acid and acid/alkaline exposure, and duodenogastric reflux. *Arch Surg* 1991 (in press).
15. DeMeester TR, Attwood SEA, Smyrk TC, Therkildsen DH, Hinder RA: Surgical therapy in Barrett's esophagus. *Ann Surg* 212:528–542, 1990.
16. Caldwell MT, Lawlor P, Byrne PJ, Walsh TN, Hennessy TP: Ambulatory oesophageal bile reflux monitoring in Barrett's oesophagus. *Br J Surg* 82:657–660, 1995.
17. Vaezi MF, Richter JE: Synergism of acid and duodenogastroesophageal reflux in complicated Barrett's esophagus. *Surgery* 117:699–704, 1995.
18. Kauer WKH, Peters JH, DeMeester TR, Ireland AP, Bremner CG, Hagen JA: Mixed reflux of gastric and duodenal juices is more harmful to the esophagus than gastric juice alone: The need for surgical therapy re-emphasized. *Ann Surg* 222(4):525–533, 1995.
19. Fein M, Ireland AP, Ritter MP, Peters JH, DeMeester TR, Bremner CG: The contribution of duodenogastric reflux to esophageal injury. *J Gastrointest Surg* 1997 (in press).

20. DeMeester TR: Prolonged oesophageal pH monitoring: In: Read NW: *Gastrointestinal Motility: Which Test?* Petersfield, England: Wrightson Biomedical Publishing Ltd, 1989:41.

21. Jamieson JR, Stein HJ, DeMeester TR, et al: Ambulatory 24-hour esophageal pH monitoring: Normal values, optimal thresholds, specificity, sensitivity, and reproducibility. *Am J Gastroenterol* 87(9):1102–1111, 1992.

22. Richter JE, Bradly LA, DeMeester TR, Wu WC. Normal 24 hour ambulatory esophageal pH values: Influence of study center, pH electrode, age and gender. *Dig Dis Sci*, 1992; 37:849–56.

# 9

# Symptom Analysis in 24-Hour Esophageal pH Monitoring

**Bas L.A.M. Weusten, M.D.**
**André J.P.M. Smout, M.D.**

By means of ambulatory 24-hour esophageal pH monitoring, the pattern and severity of gastroesophageal reflux can be quantified. The test is considered to be the gold standard for the assessment of gastroesophageal reflux in clinical practice. Uniformly, the time with an esophageal pH less than 4.0 is accepted as the most useful parameter. Based on comparison with normal values, a patient's pH profile can be classified as physiological (time with pH $< 4$ within the normal range) or as pathological (time with pH $< 4$ outside normal range). This classification, however, is not appropriate in all cases. It is well known, for instance, that subjects with abundant gastroesophageal reflux can be completely asymptomatic. In addition, data accumulates that also patients with a physiological reflux profile can suffer from reflux-induced symptoms.[1] In these patients with "endoscopy-negative reflux disease", the esophagus appears to be hypersensitive to gastroesophageal reflux (GER).[2] Ambulatory 24-hour pH monitoring is often essential to the diagnosis of this form of the disease. A firmly established diagnosis is usually required before patients with symptomatic GER are put on long-term drug treatment, or are subjected to surgery.

In order to use the test as a "physiologic Bernstein test," scrutiny of pH tracings for associations, in time, between reflux events and symptom episodes is essential. Recently, several indices have been developed to quantify symptom-reflux associations. In this chapter, the various proposed methods for symptom analysis of 24-hour esophageal pH monitoring are discussed. First, however, the question when to consider a single symptom episode as related to reflux should be answered.

# WHEN TO CONSIDER A SINGLE REFLUX EPISODE AS RELATED TO REFLUX?

## Identification of Reflux Episodes

In order to assess associations in time between reflux and symptoms, the first step is to identify episodes of reflux. Generally, a drop in pH to a value below 4 is considered to be indicative of the beginning of a reflux episode, although higher as well as lower pH thresholds have also been advocated. The threshold of pH 4 is arbitrary, but appears to be sufficient.[3] Its wide-spread use allows comparison of results between different centers. Debate exists as to whether or not a sudden drop in pH (e.g., 1 pH unit within 8 seconds) should be considered as indicative of reflux in the symptom analysis. It is certain that the esophageal pH does not necessarily have to reach values below 4 to initiate symptoms,[4] especially when the pH monitoring is performed in patients under acid suppression therapy. We and others therefore *include* these sudden drops in pH in the symptom analysis of 24-hour esophageal pH data.

## Recording of Symptom Episodes

In addition to the identification of reflux episodes, it is important to record the onset of symptoms as accurately as possible. Commercially available data loggers are therefore provided with event marker buttons. Patients should be instructed to push the event button at the moment of the occurrence of symptoms. In addition, patients can be instructed to record the time of symptom occurrence in a diary. In our laboratory, the symptom diary is only used as a check for the appropriate use of the event button. When discrepancies between the diary and the event markers are encountered, the list can be edited.

## Time Window

In those cases where symptoms are provoked by reflux, it is obvious that symptom episodes should be preceded by reflux episodes (Figure 9.1). Early studies defined a symptom as secondary to reflux when the esophageal pH dropped below 4.0 in the time window from

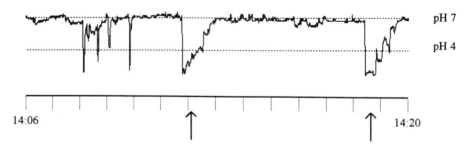

**Figure 9.1.** Part of a 24-hour esophageal pH tracing (12 minutes), showing several reflux episodes. In this example, two reflux episodes are followed by symptoms, as indicated by the arrows. The other reflux episodes remained unnoticed by the patient. Note that the symptoms in this example are preceded by reflux within a time window of 2 minutes (see text).

5 minutes before the onset of the symptom episode until its onset.[5,6] However, shorter as well as longer time windows have been used.[7,8] Johnsson and associates have shown in a group with typical reflux symptoms, that the vast majority of the symptoms occurred within the first 5 minutes after a reflux episode.[9] During the rest of the hour after a reflux episode, a more random distribution was seen. In a group of patients with noncardiac chest pain, Lam and coworkers performed repetitive symptom analysis using time windows of various onset and duration, the time windows ranging from 6 minutes before the onset of the symptom to 6 minutes after the onset of the symptom.[10] Using Poisson's theory, they concluded that the optimal time window begins at 2 minutes before the onset of a symptom episode and ends at its onset. This time window is now used by most experts working in this area.

# HOW TO EXPRESS SYMPTOM-REFLUX ASSOCIATIONS IN A SINGLE INDEX

## The Symptom Index (SI)

In 1986, Ward and colleagues were the first to introduce an index to express the strength of associations between symptoms and reflux during 24-hour esophageal pH monitoring.[5] This index, the Symptom Index (SI), was defined as:

$$\text{Symptom Index} = \frac{\text{number of reflux} - \text{related symptom episodes}}{\text{total number of symptom episodes}} \times 100\% \qquad (1)$$

A subsequent study, from the same group of investigators, examined the value of this index in 100 patients with heartburn or chest pain who underwent prolonged esophageal pH monitoring.[6] They found an excellent association between the SI and the quantitative reflux parameters: more than 90% of the patients with a high SI (greater than 75%) had abnormal quantitative reflux parameters, whereas over 70% of patients with a low SI (less than 25%) had normal quantitative reflux parameters. We recently investigated a group of 193 consecutive patients referred for ambulatory 24-hour esophageal pressure and pH monitoring.[11] Symptoms of heartburn, acid regurgitation, chest pain, and dysphagia were experienced by 125 patients during the monitoring. Pathological reflux was encountered in 34% of the patients. In the patients with pathological reflux, a high SI (50% or more) was observed in 56%, whereas in only 16% of patients with a normal pH profile a high SI was seen (Figure 9.2). Although the SI was significantly higher in patients with pathological reflux than in those without, a considerable number of patients with pathological reflux had a SI < 50%, and 16% of the patients with a normal pH profile had a SI ≥ 50% (Figure 9.2).

In the initial studies, a Symptom Index of 75% or more was (arbitrarily) considered to be significant. In other words, in a case that (at least) three out of four symptom episodes were associated with reflux, the symptoms were considered to be due to GER.[6] In later studies, the threshold of 75% was considered to be too strict,[7-9] and currently a threshold of 50% is widely used. Based on receiver operating characteristic analysis, there is evidence to suggest that an SI of 50% or more has the highest specificity and sensitivity in patients with heartburn.[12]

The major advantage of the SI is its simplicity. The SI has several important shortcomings, however, making its translation into clinical perspective more difficult than appears

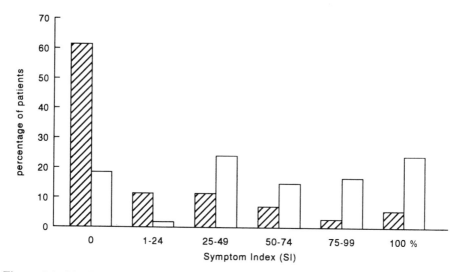

**Figure 9.2.** Distribution of the Symptom Index in patients undergoing ambulatory 24-hour esophageal pH monitoring ($n = 125$). The patients are divided in those with nonpathological (dashed bars) and those with pathological reflux (white bars) during the test. A SI $\geq 50\%$ was seen in 56% of patients with a pathological pH profile, and in 16% of those without pathological reflux ($P < 0.001$). Note that a considerable proportion of patients with a pathological pH profile had a low SI, and that a high SI was also encountered in patients with nonpathological reflux.

at first sight. First, as mentioned above, the cut-off between a positive and a negative SI is arbitrary. Second the disadvantage of the SI is that it is expressed as a percentage. An SI of 100%, for instance, simply means that all symptom episodes that occur during the 24-hour pH monitoring are, in time, associated with reflux. It can express a situation of one reflux-related symptom episode, as well as one of dozens of reflux-related symptom episodes, the latter situation being much more convincing of reflux as a cause of the symptoms than the former. Third, the SI does not take into account the total number of reflux episodes. As a consequence, the chance that a symptom is found to be associated in time with reflux is higher with increasing numbers of reflux episodes. Thus, in a patient with frequent episodes of GER who reports only one symptom episode during the 24-hour esophageal pH monitoring, a SI of 100% might well occur by chance.

## The Symptom Sensitivity Index (SSI)

In an attempt to overcome the later shortcoming of the SI, Breumelhof et al. introduced an additional parameter, the Symptom Sensitivity Index (SSI).[13] This index is defined as:

$$\text{Symptom Sensitivity Index (SSI)} = \frac{\text{number of symptomatic reflux episodes}}{\text{total number of reflux episodes}} \times 100\%$$

The index expresses the percentage of symptom-associated reflux episodes, and, quantifies the patient's sensitivity to reflux. The combined use of the SI and the SSI would provide a more accurate answer to the question whether or not the symptoms are due to reflux.

*Example 1: A patient undergoing 24-hour esophageal pH monitoring experiences*
*one symptom episode during the measurement. Analysis of the 24-hour*
*pH tracing reveals 80 reflux episodes, one of which precedes the onset*
*of the symptom episode within 2 minutes. The symptom episode is there-*
*fore considered as reflux-related. The SI = 1 ÷ 1 × 100% = 100%.*
*The SSI = 1 ÷ 80 × 100% = 1.25%.*

Similar to the SI, the SSI suffers from the disadvantage that the cut-off between a positive and a negative SSI is arbitrary. In their paper, Breumelhof et al. used a threshold of 10%. Obviously, the use of two different parameters is suboptimal. Especially, the results of patients with a discordance between the SI and the SSI (a high SI and a low SSI, and vise versa) are difficult to interpret.

## The Binomial Formula and the Kolmogorov-Smirnov Test

The drawback of the Symptom Index, that a positive SI in patients with frequent gastroesophageal reflux could well reflect random symptom-reflux associations, led Ghillebert et al. to the introduction of the binomial formula in the symptom analysis.[14] They defined symptom episodes as reflux-associated when symptoms occurred during or within two minutes of an intraesophageal drop in pH below 4, provided that the probability of this association in time was less than 5%. The first step consisted of the calculation of the probability $p$:

$$p = \frac{\text{total time (min) pH} < 4 + (2 \text{ min} \times \text{nr of reflux episodes})}{\text{total time (min) of recording}}$$

The probability that observed associations between symptom and reflux during 24-hour pH monitoring occurred only by chance was obtained by substitution of $p$ in the binomial formula:

$$\frac{n!}{r!\,(n-r)!}\, p^r\,(1-p)^{n-r}$$

in which $n$ is the total number of symptom episodes, and $r$ ranges from the number of reflux-related symptom episodes to $n$. In contrast to the Symptom Index, the total number of reflux episodes is taken into account in the binomial formula. The calculation of the binomial formula, however, is rather complicated. In addition, there is one important shortcoming in this method. In the first step of the calculation (the quotient in which the probability $p$ is calculated), overlap between successive reflux episodes occurs when a reflux episode occurs within 2 minutes after another. This results in an artificially high numerator (Figure 9.3). This makes the binomial formula unreliable, especially in patients with frequent episodes of gastroesophageal reflux.

Armstrong et al. proposed another method, using the Kolmogorov-Smirnov test.[15] Basically, the technique tests whether the distribution of pH values recorded during the symptom episodes is statistically different from the distribution of pH values in the asymptomatic episodes. The method suffers from the disadvantage that an incorrect time window is used, and that one-sided testing is impossible.

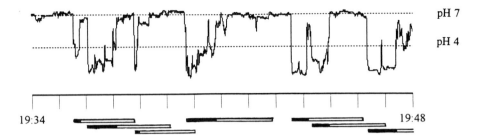

**Figure 9.3.** Part of a pH tracing showing multiple reflux episodes in rapid succession. In this example, the drawback of the binomial formula according to Ghillebert is illustrated (see text). The horizontal bars at the bottom side indicate the time with pH < 4 + 2 minutes. Note the overlap in bars, indicating the discongruency between the calculated time and the actual time in which an occurring symptom episode would be considered to be associated with reflux. (From: Weusten BLAM, Akkermans LMA, vanBerge-Henegouwen GP, et al.: Ambulatory combined esophageal pressure and pH monitoring: relationships between pathological reflux, esophageal dysmotility and symptoms of esophageal dysfunction. *Eur J Gastroenterol Hepatol* 5:1055–1060, 1993 permission obtained).

## The Symptom Association Probability (SAP)

We recently developed a simple statistical method resulting in a single parameter expressing the probability that observed associations in time between reflux and symptoms are *not* caused by chance: the Symptom Association Probability (SAP).[16] In the calculation of the SAP, the optimal 2 min time window is used. The calculation procedure consists of the following steps. First, the 24 hr esophageal pH signal is divided into consecutive 2 min periods. All of these are evaluated for the occurrence of reflux and hence classified as either reflux-positive or as reflux-negative. Then, the 2 min periods preceding the onset of symptom episodes are analyzed for the presence of reflux and classified accordingly. Using the thus acquired data, a contingency table can be constructed, containing 4 fields: one field containing the number of symptomatic reflux-positive 2-min periods ($S^+R^+$), one field with the number of asymptomatic reflux-positive 2 min periods ($S^-R^+$), one with symptomatic 2 min periods without reflux events ($S^+R^-$), and one field with the number of asymptomatic 2 min periods without reflux events ($S^-R^-$):

|  | Symptoms | | |
|---|---|---|---|
|  | + | − |  |
| R e f **+** | $S^+R^+$ | $S^-R^+$ | $R^+$ |
| l u **−** | $S^+R^-$ | $S^-R^-$ | $R^-$ |
| x | $S^+$ | $S^-$ | *Total* |

When the symptoms are completely unrelated to the occurrence of reflux events, the *symptomatic* 2 min periods should be distributed proportionally over the reflux-positive and reflux-negative 2 min periods. When reflux provokes the symptoms, one would expect the proportion of reflux-positive symptomatic 2 min periods to be higher than that expected by chance. Using the Fisher Exact test, the probability (P value) can be calculated as the observed association between reflux and symptoms occurred by chance.[17] The Symptom Association Probability is calculated as $(1.0 - P) \times 100\%$.

The first advantage of the SAP is that it takes all relevant data into account, leading to a single parameter expressing the probability that observed symptom-reflux associations are not caused by chance, in other words, expressing the probability that symptoms and reflux are associated. Second, the problem of the arbitrary threshold between a positive and a negative parameter is solved: the generally accepted statistical border of 95% confidence interval is used. Third SAP makes use of the optimal 2 min time window between the onset of reflux episodes and the occurrence of symptoms. The calculation of the SAP can easily be integrated in the existing 24-hour esophageal pH tracing analysis software.

*Example 2: A patient undergoing 24 hour esophageal pH monitoring experiences 4 symptom episodes during the monitoring. Two of these are preceded within 2 minutes by a reflux episode. Division of the entire pH profile into consecutive 2-min periods results in 720 2-min periods, and 35 of them show evidence of GER. The following contingency table can be constructed:*

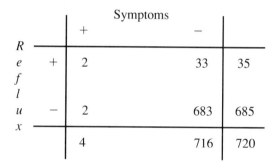

*The Fisher's exact test, expressing the probability that the found distribution of the symptomatic 2-min periods among the reflux-positive and the reflux-negative 2-min periods is caused by chance, results in a P-value of 0.0126. The SAP is $[1-0.0126] \times 100\% = 98.7\%$. Hence, the probability that the symptom-reflux association in this patient is not caused by chance is more than 95%, rendering reflux as a cause of the symptoms in this patient highly likely.*

# CONCLUSION

In the analysis of 24-hour esophageal pH data, symptom analysis is of crucial importance. Symptom analysis and quantitative 24-hour pH monitoring results complement each other, giving the physician more valid information on which to base therapy. When there is only a weak association in time between symptoms and the occurrence of reflux, or no association at all, the patient can be reassured that gastroesophageal reflux is not the cause of the patient's complaints. On the other hand, when the endoscopic examination is normal, a high symptom-reflux association in conjunction with an abnormal 24-hour pH test enables the physician to be confident that the symptoms are directly related to gastroesophageal reflux. In those patients in whom the pH test results and the results of symptom analysis are contradictory, it is important to review the pH tracing for equipment failure and to make sure that the patient did not markedly alter his or her activity or diet (e.g., could not eat regularly with the probe inserted) during the test. If these factors can be excluded, the symptom analysis identifies patients who would be improperly classified by quantitative 24-hr pH testing alone.

Several methods for symptom analysis have been developed, all with their own restrictions. We recently compared the results of symptom analysis by means of the calculation of the SI with that of the SAP in 184 patients with at least one symptom during 24-hour esophageal pH monitoring.[16] Discordance between SI and SAP values occurred in 11% of cases. In accordance with what one could expect from the above mentioned drawbacks of SI, the false-positive and false-negative SI values occurred preferentially in patients with small and large numbers of symptom episodes during the test, respectively. Other studies comparing the various symptom analysis techniques are lacking. Most important, no data are available that would allow one to decide which type of symptom analysis best predicts the outcome of treatment in GERD.* Therefore, the choice of which index to use can only be based on theoretical grounds. We believe that the symptom association probability currently is the best available method, taking all relevant data into account.

# REFERENCES

1. Vantrappen G, Janssens J, Ghillebert G: The irritable oesophagus — a frequent cause of angina-like pain. *Lancet* ii: 1232–1234, 1987.
2. Trimble KC, Pryde A, Heading RC: Lowered esophageal sensory thresholds in patients with symptomatic but not excess gastro-oesophageal reflux: evidence for a spectrum of visceral sensitivity in GERD. *Gut* 37:7–12, 1995.
3. Weusten BLAM, Roelofs JMM, Akkermans LMA, vanBerge-Henegouwen GP, Smout AJPM: Objective determination of pH thresholds in the analysis of 24-hour ambulatory esophageal pH monitoring. *Eur J Clin Invest* 26:151–158, 1996.
4. Breumelhof R, Nadorp JHSM, Akkermans LMA, Smout AJPM: Analysis of 24-hour esophageal pressure and pH data in unselected patients with noncardiac chest pain. *Gastroenterology* 99:1257–1264, 1990.
5. Ward BW, Wu WC, Richter JE, Lui KW, Castell DO: Ambulatory 24-hour esophageal pH monitoring. Technology searching for a clinical application. *J Clin Gastroenterol* 8:59–67, 1986.
6. Wiener GJ, Richter JE, Copper JB, Wu WC, Castell DO: The symptom index: a clinically important parameter of ambulatory 24-hour esophageal pH monitoring. *Am J Gastroenterol* 83:358–361, 1988.
7. Peters L, Maas L, Petty D, et al: Spontaneous noncardiac chest pain. Evaluation by 24-hour ambulatory esophageal motility and pH monitoring. *Gastroenterology* 94:878–886, 1988.

*Editor's Note: To date, no large study has compared the predictive accuracy of these methods for symptom analysis in a patient population with heartburn where acid reflux was aggressively treated and control documented by prolonged esophageal pH monitoring. This important study is eagerly awaited.

8. Baldi F, Ferrarini F, Longanesi A, Ragazzini M, Barbara L: Acid gastroesophageal reflux and symptom occurrence. Analysis of some factors influencing their association. *Dig Dis Sci* 34:1890–1893, 1989.

9. Johnsson F, Joelsson B, Gudmundsson K, Greiff L: Symptoms and endoscopic findings in the diagnosis of gastroesophageal reflux disease. *Scand J Gastroenterol* 22:714–718, 1987.

10. Lam HGT, Breumelhof R, Roelofs JMM, vanBerge-Henegouwen GP, Smout AJPM: What is the optimal time window in symptom analysis of 24-hour esophageal pressure and pH data? *Dig Dis Sci* 39:402–409, 1994.

11. Weusten BLAM, Akkermans LMA, vanBerge-Henegouwen GP, Smout AJPM: Ambulatory combined oesophageal pressure and pH monitoring: relationships between pathological reflux, oesophageal dysmotility and symptoms of oesophageal dysfunction. *Eur J Gastroenterol Hepatol* 5:1055–1060, 1993.

12. Singh S, Richter JE, Bradley LA, Haile JM: The symptom index. Differential usefulness in suspected acid-related complaints of heartburn and chest pain. *Dig Dis Sci* 38:1402–1408, 1993.

13. Breumelhof R, Smout AJPM: The symptom sensitivity index: a valuable additional parameter in 24-hour esophageal pH recording. *Am J Gastroenterol* 86:160–164, 1991.

14. Ghillebert G, Janssens J, Vantrappen G, Nevens F, Piessens J: Ambulatory 24-hour intra-oesophageal pH and pressure recordings v provocation tests in the diagnosis of chest pain of oesophageal origin. *Gut* 31:738–744, 1990.

15. Armstrong D, Emde C, Inauen W, Blum AL: Diagnostic assessment of gastroesophageal reflux disease: what is possible vs. what is practical? *Hepato-gastroenterol* 39:3–13, 1992.

16. Weusten BLAM, Roelofs JMM, Akkermans LMA, vanBerge-Henegouwen GP, Smout AJPM: The Symptom Association Probability: an improved method for symptom analysis of 24-h esophageal pH data. *Gastroenterology* 107:1741–1745, 1994.

17. Lentner C, Diem K, Seldrup J, eds: Geigy Scientific Tables. 8th ed. v. 2. Basle: Ciba-Geigy, p227, 1982.

# 10

# pH Monitoring Versus Other Tests for Gastroesophageal Reflux Disease: Is This the Gold Standard?

**Donald O. Castell, M.D.**

Over the last 20 years there has been increasing awareness of the clinical importance of gastroesophageal reflux disease and of techniques to accurately diagnose this condition. The diagnostic approach to the patient with possible reflux esophagitis is often confusing, leaving the clinician with many questions regarding the preferred course of action. Since the symptoms are primarily attributable to chronic gastroesophageal reflux, the intent of the evaluation is to document this abnormality. For the patient who has typical heartburn and regurgitation with the usual positional and postprandial relationships, little, if any, additional information is required to establish a presumptive diagnosis and initiate therapy. If the patient's symptoms are less clear-cut or include any of the various extraesophageal or atypical manifestations, additional diagnostic testing is usually appropriate.

A variety of tests and procedures are available to evaluate the status of a patient with definite or possible gastroesophageal reflux disease. Clarifying which is the preferred or indicated mode of testing should be approached initially by defining the diagnostic question to be answered in the individual patient. Most simply stated, if evidence of esophagitis or Barrett's esophagus is being sought, endoscopy is the preferred test. If, on the other hand, one wishes to identify the degree of abnormal reflux and/or its relation to symptoms, ambulatory pH monitoring will provide the most information. Table 10.1 lists the variety of diagnostic tests that have been advocated for patients with possible gastroesophageal reflux and categorizes them according to the diagnostic question.

**TABLE 10.1. CATEGORIES OF DIAGNOSTIC TESTS AVAILABLE TO ANSWER SPECIFIC QUESTIONS ABOUT POSSIBLE GASTROESOPHAGEAL REFLUX DISEASE**

*Is abnormal reflux present?*
  Barium upper gastrointestinal series
  Gastrophageal scintiscan
  Standard acid reflux test
  Prolonged ambulatory pH monitor
*Is there reflux injury?*
  Barium upper gastrointestinal series (air contrast)
  Endoscopy
  Mucosal biopsy
  Potential difference measurement
*Are symptoms due to reflux?*
  Bernstein's test
  pH monitor (with symptom index)
*Prognostic or preoperative assessment*
  Esophageal manometry (both lower esophageal sphincter and peristalsis)
  pH monitor

# DOES THE PATIENT HAVE ABNORMAL GASTROESOPHAGEAL REFLUX?

Whether or not the patient has abnormal gastroesophageal reflux is the critical question frequently asked to confirm the diagnosis in patients with the more typical symptom pattern of heartburn and/or positional regurgitation, to help clarify the diagnosis in patients with atypical or extraesophageal symptoms, and to evaluate the effectiveness of acid suppression in those whose typical heartburn symptoms have not responded to appropriate initial therapy. Simply confirming that abnormal reflux is present supports the clinical impression.[1]

The first test often performed to evaluate possible reflux, particularly by the primary care physician, is a *barium upper gastrointestinal series*. This test may be particularly helpful if the patient has severe or persistent gastroesophageal reflux and will also help to rule out complications of gastroesophageal reflux disease (ulcer; stricture but not Barrett's esophagus) or other diagnosis, e.g., structural abnormalities of the esophagus, stomach, or duodenum. Although the barium study may be useful, it is important to recognize its limitations. Radiographic reflux, the flow of barium from the stomach into the esophagus either spontaneously or induced by various maneuvers (e.g., Trendelenburg's position; abdominal compression; water siphon), is an objective sign of gastroesophageal reflux, but is limited by the state of competence of the antireflux mechanism at a particular moment. Based on our comparison of results of 125 patients having both upper gastrointestinal series and ambulatory pH monitoring, we find the barium study to be a poor predictor of abnormal reflux.[2] Another recent report, however, has suggested that a standardized barium study, including water syphon, will yield more reliable results with a positive predictive value of 80%.[3]

Radioisotope scintigraphy has been employed to document reflux and also to provide a quantitative assessment of reflux. This *gastroesophageal scintiscan* employs a radioisotope ($^{99m}$Tc-sulfur colloid, 100 μCi) as a marker for reflux. In addition, increasing abdominal compression is used to unmask incompetence of the reflux barrier. Initial experience with the gastroesophageal scintiscan suggested that it had high specificity and sensitivity for reflux.[4] More recent results have shown a sensitivity of only 60 percent but confirmed the high specificity.[5]

The *standard acid reflux test* was developed to stress the antireflux barrier while measuring possible reflux with a pH electrode placed 5 cm above the lower esophageal sphincter. After 300 ml of 0.1 normal hydrochloric acid is instilled into the stomach, the patient performs four maneuvers: deep breathing, Valsalva, Mueller (inspiration against a closed glottis), and cough. All four maneuvers are repeated in the supine, right, and left lateral decubitus positions and with the head down 20 degrees. Overall, 16 possibilities for acid reflux may occur. A decrease in esophageal pH to less than 4.0 on at least three occasions is considered evidence of abnormal gastroesophageal reflux. This test has an overall sensitivity and specificity of about 80 percent, although a false positive rate as high as 31 percent has been reported.[6] It has been essentially replaced by the more physiologic technique of prolonged pH monitoring.

*Prolonged ambulatory esophageal pH monitoring* is the preferred method for detecting the presence of abnormal reflux, since it measures the actual exposure time of the distal esophageal mucosa to acid gastric juice. This test records reflux during normal daily activities such as upright activity, eating, and sleeping or lying recumbent. It does not stress the lower esophageal sphincter with unnatural respiratory maneuvers or unphysiologic abdominal compression.

Ambulatory pH monitoring is the most physiologic test currently available for studying reflux, and frequent refinements in technique allow accurate recording over prolonged time intervals. It is generally accepted as the gold standard for identification of reflux. However, it is important to understand the limitations of this technique and to accept that the findings are not absolute. In the study by Schlesinger and colleagues, a group of 64 hospitalized patients with typical reflux symptoms were compared with 20 age-matched controls.[7] The patients were divided into two groups: 30 with reflux symptoms and normal endoscopic examination (group 1) and 34 with similar symptoms and erosive esophagitis (group 2). As in prior studies, the patients with typical reflux symptoms had significantly more reflux than did the controls, but only 48 percent in group 1 had abnormal results. Furthermore, only 71 percent of patients with erosive esophagitis (group 2) had an abnormal 24 hour pH monitoring result. These results, shown in Figure 10.1, suggest that esophageal pH monitoring lacks sensitivity or accuracy and that a negative test can be misleading. This may be especially true in hospitalized patients where activity levels and habits may differ greatly from their home and work environment.

Klauser and coworkers report similar observations finding that 17 percent of patients with endoscopic esophagitis had normal pH testing.[8] In addition, only 72 percent of patients characterized as having "definite" gastroesophageal reflux by symptom assessment were found to have abnormal pH monitoring. In the latter situation, in which the presence of gastroesophageal reflux disease is based on typical reflux symptoms, no true criterion is available to serve as a basis for comparative diagnosis. Much of the information in the literature comparing the sensitivity and specificity of the various tests for gastroesophageal reflux disease has utilized such symptomatic assessment. In addition, since there is ample documentation that reflux can occur in the absence of esophagitis, the en-

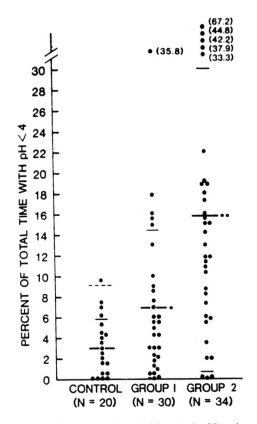

**Figure 10.1.** Total esophageal acid exposure time in 20 controls, 30 patients with reflux symptoms and normal endoscopy (group 1), and 34 patients with reflux symptoms and erosive esophagitis (group 2). Two standard deviations above the mean for controls is shown(----).

doscopic assessment does not provide a reliable absolute diagnostic criterion. It is apparent that occasional false negative results of ambulatory pH monitoring may occur but it seems unlikely that false positive results occur.

Observations such as the preceding underscore the difficulty in placing numerical values for true sensitivity and specificity for any test of gastroesophageal reflux disease. In order to define these characteristics for any test, valid criteria discriminating between the presence and absence of the disease in question must be available. The presence of erosive esophagitis on endoscopic examination provides such strong evidence for the presence of gastroesophageal reflux disease that one must accept the potential for pH monitoring to provide false negative results. Possible explanations for this finding are discussed below. The relatively low frequency of abnormal pH studies in patients with symptoms suggestive of reflux provides considerably weaker evidence. Because symptoms suggestive of reflux may be produced by a variety of other causes, these findings do not provide a meaningful indictment against the status of pH monitoring as a gold standard.

As noted above, discrepancies between pH monitoring and documented endoscopic esophagitis support the conclusion that false negative pH monitoring studies do occur. An explanation for this phenomenon is supported by studies from our laboratory in which two pH probes were placed at the same level (5 cm above the lower esophageal sphincter) in a group of patients with reflux.[9] Considerable discrepancies were noted between simultaneous readings from the two probes in the same patient, as shown in Figures 10.2A and

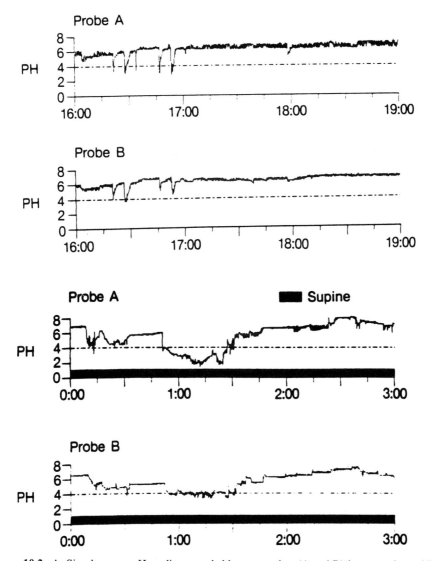

**Figure 10.2.** A, Simultaneous pH studies recorded by two probes (A and B) in one patient while in the upright position. Five short duration episodes are recorded by probe A, whereas only one such episode is recorded by probe B. B, Simultaneous pH studies recorded by two probes (A and B) in a patient while in the supine position. A reflux episode of long duration occurred at 12:50, as recorded by probe A, whereas probe B recorded multiple short episodes during a similar time period.

10.2B. These differences would have resulted in a change in diagnosis (normal versus abnormal) in 20 percent of the patients, using time below pH 4.0 as the critical factor. This effect was hypothesized to be due to the potential for the pH electrode at the tip of the probe to be buried within the esophageal mucosa and actually miss adjacent acid. Other studies from our laboratory with two ambulatory pH tests performed within 10 days yielded comparable results.[10] When patients and controls were compared, similar variability was noted in the same subjects between the two days of testing, with the pH monitoring results producing a change in diagnosis (normal or abnormal based on percentage of the time the pH was below 4.0) in 11 percent of cases (Figure 10.3). Evaluated from the opposite view, these data also indicate that the same diagnosis was obtained in 89 per-

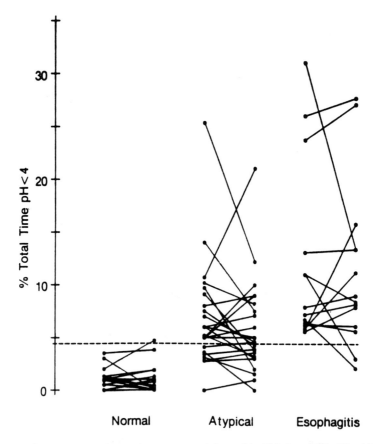

**Figure 10.3.** Absolute values for total percentage of time with pH below 4.0 in 53 subjects studied by ambulatory 24 hour esophageal pH monitoring performed on two separate days. The total group can be divided into three subgroups representing the clinical spectrum of gastroesophageal reflux disease: normal volunteers, patients with atypical reflux symptoms (e.g., chest pain, asthma), and patients with esophagitis. The dotted line represents the $\bar{X} + 2$ standard deviations for total percentage of time with the pH below 4.0 derived from 20 asymptomatic volunteers (i.e., the diagnostic discrimination value for abnormal gastroesophageal reflux).

cent of the cases. This study was also dependent on the variability in the amount of reflux that occurs in any one individual from day to day. The observation that 7 percent of our patients with endoscopic esophagitis had at least one entirely normal pH study underscores the fact that even pH monitoring is not an absolute discriminator for gastroesophageal reflux disease.

It seems reasonable to conclude that the ambulatory pH monitor may not be a true gold standard but that it is reasonably reliable as a diagnostic test once its limitations are understood. It is the best test to document the presence of abnormal reflux. In addition, identifying the specific association of symptoms, particularly chest pain, cough, or asthma, with reflux episodes increases the diagnostic usefulness of pH monitoring.[11,12]

# IS THERE EVIDENCE OF ESOPHAGEAL INJURY SECONDARY TO GASTROESOPHAGEAL REFLUX?

In many patients it is preferable to identify evidence of esophagitis in order to confirm the diagnosis and to establish the extent of gastroesophageal reflux disease present. Usually this requires consultation with a gastroenterologist to perform endoscopy. An exception is obtained via use of a carefully performed *air contrast barium esophagram*. By incorporating the air contrast technique into the barium study of the esophagus, mucosal lesions (erosions or ulcers) or strictures may be identified. Comparison of this technique with subsequent endoscopic examination has revealed that the finding of esophageal injury on air contrast esophagram is quite specific although not very sensitive.[13] Thus, if this simple screening test is positive for esophagitis, there may be no need to proceed to the more specialized techniques described below. If the esophagram is negative, more sensitive testing is required.

As mentioned above, *endoscopy* is usually the diagnostic approach utilized to document esophageal injury. Erosions or ulceration of the esophageal mucosa visualized through the endoscope are indications of injury resulting from gastroesophageal reflux once infectious esophagitis or medication-induced injury is excluded. The findings are definitive when clearly present, but unfortunately they may be quite subtle or absent, even in the presence of specific histologic abnormalities on a mucosal biopsy specimen.

*Esophageal mucosal biopsy* provides a more sensitive test for the presence of esophageal injury secondary to gastroesophageal reflux because histologic abnormalities may be present even when careful endoscopic examination indicates a normal-appearing esophagus. The most reliable criterion for esophagitis on endoscopic biopsy is the presence of acute inflammatory cells (polymorphonuclear leukocytes or eosinophils). Polymorphonuclear leukocytes are seen in the esophageal biopsy specimens of only about 20 percent of patients with gastroesophageal reflux symptoms. Therefore, other epithelial changes have been proposed as more sensitive criteria for the histologic diagnosis of reflux disease, particularly increased papillary extension and basal zone hyperplasia.[14] Biopsy also is important to confirm a diagnosis of Barrett's esophagus and to evaluate possible dysplasia.

# ARE THE PATIENT'S SYMPTOMS DUE TO GASTROESOPHAGEAL REFLUX?

The key question in many patients is to demonstrate that their symptoms are clearly secondary to acid sensitivity of the esophageal mucosa due to chronic gastroesophageal reflux. The *acid perfusion (Bernstein) test* has been used for many years as a test of acid sensitivity. The reported specificity and sensitivity are both approximately 80 percent in gastroesophageal reflux disease.[1] If the patient's symptoms are reproduced during perfusion of dilute hydrochloric acid and clear following saline perfusion, it is appropriate to conclude that chronic acid reflux is the cause of spontaneously occurring symptoms. This test, of course, is purely qualitative in nature and provides no information on the degree of gastroesophageal reflux.

*Prolonged ambulatory pH monitoring* can provide more accurate information on the relationship of specific symptoms to reflux if the symptom index is calculated during the period of testing. This technique, which may be considered an endogenous Bernstein test, is fully discussed in Chapter 9.

# TESTING TO PROVIDE PROGNOSTIC OR PREOPERATIVE ASSESSMENT

Measurement of *lower esophageal sphincter pressure* as a possible means for diagnosing the presence of reflux disease is supported by recent studies emphasizing its importance as a major barrier to reflux. Although a resting lower esophageal sphincter pressure of less than 10 mm Hg has been considered typical of patients with an incompetent esophagogastric junction, there is much variation in this value. Many patients with well-documented esophagitis have lower esophageal sphincter pressures greater than 10 mm Hg and occasional asymptomatic subjects have pressures below this value. This observation is not surprising considering the current belief that transient relaxations of the sphincter are the primary mechanisms by which reflux occurs.[15] Consequently, lower esophageal sphincter pressure in a given patient is too imprecise for identifying a potential for reflux, even though it distinguishes populations of symptomatic patients from controls.[16] Lower esophageal sphincter variability severely limits the sensitivity of its measurement as a diagnostic test. A pressure of less than 6 mm Hg shows reasonably high specificity compared with abnormal reflux on pH testing and may have important predictive value in identifying a more severe degree of reflux and a poorer prognosis for long-term medical therapy.[17]

Quantitative assessment of *peristaltic activity* in the esophageal body is also an important test in assessing the severity of the reflux disease and its prognosis.[18] This test is particularly important as a preoperative assessment to inform the surgeon of potentially defective peristalsis prior to performing fundoplication. In our laboratory, low amplitude (or ineffective) peristalsis in the distal esophagus is more prevalent (34 percent) in patients with reflux disease than is a sphincter pressure < 10 mm Hg (4 percent).

As a preoperative evaluation, ambulatory pH monitoring can provide important information about the severity of reflux and the reflux pattern present in a particular patient, such as does the patient have reflux predominantly at night or is upright, postprandial reflux more prevalent?

# IS THE FINDING OF
# A HIATUS HERNIA IMPORTANT?

Although once considered to be of major importance in the production of gastro-esophageal reflux, the finding of a hiatus hernia either radiographically or during endoscopic evaluation has little importance in predicting whether a patient's symptoms are secondary to reflux. When carefully sought, a sliding hiatus hernia can be found in a high percentage of individuals, most of whom are asymptomatic. In addition, recent studies have indicated no definite cause and effect relationship between the presence of symptomatic gastroesophageal reflux and the finding of a hiatus hernia.[19] Since both heartburn and a sliding hiatus hernia are common phenomena, association between these entities is not unexpected. Such an observation does not necessarily imply a cause and effect relationship. Recent information has indicated that hiatus hernias may contribute to the severity of gastroesophageal reflux by "trapping" acid in the distal esophagus and delaying its clearance from the esophageal mucosa.[20] Thus, although the hernia may not necessarily produce the gastroesophageal reflux, it may contribute to the esophageal injury by prolongation of acid exposure. In addition, recent studies have indicated that hiatus hernias may not be present in most patients with milder forms of gastroesophageal reflux disease but that the vast majority of patients with severe esophagitis are likely to have a hiatus hernia discovered during endoscopic assessment.[19]

# AN INTEGRATED APPROACH TO THE DIAGNOSIS
# OF GASTROESOPHAGEAL REFLUX DISEASE

It is important to recognize that all the tests and procedures available for investigating the patient with reflux esophagitis are not needed in the individual patient. Depending on the status of the patient and the question to be answered, the proper tests can be selected. A variety of tests and procedures have been developed, as shown in Table 10.1. In actual practice only four of these are routinely utilized barium upper gastrointestinal series, endoscopy, ambulatory pH monitoring, and manometry. Figures 10.4 and 10.5 offer suggested algorithms for the approach to the patient with possible gastroesophageal reflux disease. The approach can be modified depending upon whether the patient is being evaluated by a gastroenterologist or a primary care physician, with a barium study replacing endoscopy. The observation that an underlying Barrett's esophagus may be found in up to 12 percent of patients with chronic reflux[21] underscores the importance of early endoscopy to identify this diagnosis. Manometry should be performed prior to anti-reflux surgery.

Clearly, when the patient has typical reflux symptoms and visible esophagitis, further confirmation of the diagnosis is not needed. However, in many patients ambulatory pH monitoring can provide extremely useful information. These include patients with equivocal reflux symptoms, with atypical symptoms potentially attributable to reflux (chest pain, hoarseness, wheezing, or dyspepsia), and patients with typical heartburn who failed to respond to appropriate therapy. Ambulatory pH monitoring can also provide preoperative evaluation of reflux severity and patterns prior to performing fundoplication, and evalua-

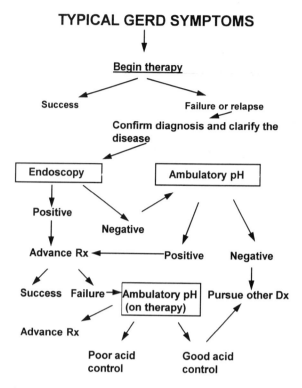

**Figure 10.4.** Algorithm for the approach to the patient with typical reflux symptoms.

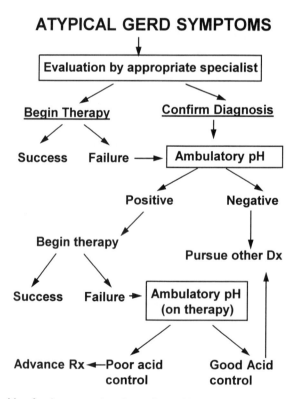

**Figure 10.5.** Algorithm for the approach to the patient with atypical reflux symptoms.

TABLE 10.2. INDICATIONS FOR PROLONGED AMBULATORY pH MONITORING IN PATIENTS WITH POSSIBLE GASTROESOPHAGEAL REFLUX DISEASE

| |
|---|
| Equivocal symptoms of gastroesophageal reflux disease |
| Atypical or extraesophageal symptoms |
|    Chest pain |
|    Hoarseness |
|    Asthma/chronic cough |
| Typical symptoms unresponsive to appropriate therapy |
| Preoperative assessment |
| Evaluation of effect of therapy possibly with probe in esophagus and stomach. |

tion of the effect of medical or surgical therapies on reflux (Table 10.2). At present, use of dual electrode (esophagus and stomach) monitoring of patients who have an inadequate response to medical therapy is becoming one of the more frequent indications for ambulatory pH monitoring.

# REFERENCES

1. Richter JE, Castell DO: Gastroesophageal reflux. Pathogenesis, diagnosis, and therapy. *Ann Intern Med* 97:93–103, 1982.
2. Johnston BT, Troshinsky MB, Castell JA, Castell DO: Comparison of barium radiology with esophageal pH monitoring in the diagnosis of gastroesophageal reflux disease. *Am J Gastroenterol* 91:1181–1185, 1996.
3. Thompson JK, Koehler RE, Richter JE: Detection of gastroesophageal reflux: Value of barium studies campared with 24-hr pH monitoring *AJR* 162:621–626, 1994.
4. Fisher RS, Malmud LS, Roberts GS, et al: Gastroesophageal (GE) scintiscanning to detect and quantitate GE reflux. *Gastroenterology* 70:301–307, 1976.
5. Jenkins AF, Cowan RJ, Richter JE: Gastroesophageal scintigraphy: Is it a sensitive screening test for gastroesophageal reflux disease? *J Clin Gastroenterol* 7:127–131, 1985.
6. Meyers WF, Roberts CC, Johnson DG, et al: Value of tests for evaluation of gastroesophageal reflux in children. *J Pediatr Surg* 20:515–520, 1985.
7. Schlesinger PK, Donahue PE, Schmid B, et al: Limitations of 24-hour intraesophageal pH monitoring in the hospital setting. *Gastroenterology* 89:797–804, 1985.
8. Klauser AG, Heinrich C, Schindlbeck NE, et al: Is long-term esophageal pH monitoring of clinical value? *Am J Gastroenterol* 84:362–365, 1989.
9. Murphy D, Yuan Y, Castell D: Does the intraesophageal pH probe accurately detect acid reflux? Simultaneous recording with two pH probes in humans. *Dig Dis Sci* 34:649–656, 1989.
10. Wiener GJ, Morgan TM, Copper JB, et al: Ambulatory 24-hour esophageal pH monitoring: Reproducibility and variability of pH parameters. *Dig Dis Sci* 33:1127–1133, 1988.
11. Wiener GJ, Richter JE, Cooper JB, Wu WC, Castell DO: The symptom index: A clinically important parameter of ambulatory 24-hour esophageal pH monitoring *Am J Gastroenterol* 83:358–361, 1988.
12. Kahrilas PJ, Quigley EM: Clinical esophageal pH recording: a technical review for practice guideline development. *Gastroenterology,* 110:1981–1996, 1996.
13. Ott DJ, Gelfand DW, Wu WC: Reflux esophagitis: Radiographic and endoscopic correlation. *Radiology* 130:583–588, 1979.

14. Ismail-Beigi F, Horton PF, Pope CE: Histological consequences of gastroesophageal reflux in man. *Gastroenterology* 58:163–174, 1970.
15. Penagini R, Schoeman M, Dent J, et al: Motor events underlying gastro-esophageal reflux in ambulant patients with reflux oesophagitis. *Neurogastroenterol Mot* 8:131–141, 1996.
16. Kraus BB, Wu WC, Castell DO: Comparison of lower esophageal sphincter manometrics and gastroesophageal reflux measured by 24-hour pH recording. *Am J Gastroenterol* 88:692–696, 1990.
17. Lieberman DA: Medical therapy for chronic reflux esophagitis: Long-term follow-up. *Arch Intern Med* 147:1717–1720, 1987.
18. Kahrilas PJ, Dodds WJ, Hogan WJ, et al: Esophageal peristaltic dysfunction in peptic esophagitis. *Gastroenterology* 91:897–904, 1986.
19. Sloan S, Rademaker AW, Kahrilas AJ: Determinants of gastroesophageal junction incompetence: hiatus hernia, lower esophageal sphincter, or both? *Ann Intern Med* 117:977–82, 1992.
20. Mittal RK, Lange RC, McCallum RW: Identification and mechanism of delayed esophageal acid clearance in subjects with hiatus hernia. *Gastroenterology* 92:130–135, 1987.
21. Winters C, Spurling T, Chobanian S, et al: Barrett's esophagus. A prevalent, occult complication of gastroesophageal reflux disease. *Gastroenterology* 92:118–124, 1987.

# 11

# Chest Pain and Gastroesophageal Reflux Disease

## Joel E. Richter, M.D.

Recurring substernal chest pain is an important clinical problem, causing anxiety for patients and their physicians because they both fear the possibility of cardiac disease. Even after cardiac disease has been excluded, many patients continue to have chest pain, persistent anxiety, and compromised lifestyles. Despite their physicians' reassurances, many patients also continue to believe that they have cardiac disease. Furthermore, this fear is perpetuated because physicians maintain these patients on medications commonly prescribed for cardiac problems, such as nitrates and calcium channel blockers.[1]

How common is noncardiac chest pain? The most recent survey in the United States estimated that approximately 600,000 new patients per year have cardiac catheterization.[2] Normal coronary arteries or "insignificant degrees of obstruction" are found in up to 30 percent of patients with anginal syndromes.[3] Thus, as many as 180,000 new cases of noncardiac chest pain are identified yearly in the United States. These figures are probably grossly underestimated because many patients do not undergo cardiac catheterization, particularly if they are under 40 years of age.

Understanding the etiology of recurrent noncardiac chest pain is a major clinical dilemma. The location of the esophagus near the heart and its similar neural pathway have made it a logical alternative explanation for these patients' chest pain complaints. Over the last 20 years, investigators have attempted to assess the importance of esophageal diseases in patients with noncardiac chest pain. However, prevalence figures are quite variable and dependent on the sophistication of esophageal studies and the type of patients studied. Overall, these series suggest that esophageal diseases may account for symptoms in 18 to 58 percent of patients with chest pain and normal coronary arteries.[3] Although esophageal motility disorders were

**119**

once believed to be the primary esophageal cause of noncardiac chest pain symptoms, recent studies now suggest that gastroesophageal reflux disease may be much more common.

## Case Study No. 1

R.J. is a 52 year old man who smokes two packs of cigarettes per day and has mildly high blood pressure. Over the last 3 months he began to notice brief episodes of substernal chest pain, generally after meals and late in the evening. These pains did not radiate, lasted a couple of minutes, and often were relieved by belching. One night, after a particularly enjoyable business dinner replete with French cooking, wines, and after-dinner liqueurs, R.J. awoke suddenly at 2 A.M. with burning substernal chest pain associated with diaphoresis and frequent belching. He walked around the room for about 30 minutes, drank several glasses of water, and took some antacid tablets, without an improvement in his symptoms. Finally, his wife drove him to the emergency room.

When seen by the emergency room physician, R.J.'s pain had resolved but he was anxious and clammy. His blood pressure was 160/110 in a sitting position with a heart rate of 95 beats per minute. There was no chest wall or abdominal distress. On further questioning, it was discovered that the patient's father had died of a myocardial infarction in his middle 50s and the patient's older brother had angina pectoris. An electrocardiogram was normal, but R.J. spent the night in the coronary care unit until cardiac enzyme tests had excluded a myocardial infarction.

Over the next several weeks, R.J. had several more episodes of substernal burning pain, which lasted from 30 minutes up to 1 hour. The symptoms usually came on after a late evening meal, while bending over working in his garden, or when he got particularly excited at work. One very troublesome episode occurred while he was playing racquetball. The patient's family physician referred him to a cardiologist. An thallium exercise study was normal. After a second admission to the coronary care unit for chest pain awakening him at night, the patient underwent coronary angiography with ergonovine testing. This study revealed normal epicardial coronary arteries without evidence of coronary spasm.

With these test results, the patient's cardiologist was reassured, but R.J. continued to ask ". . . but if it isn't my heart, what is the cause of my pains?" The patient was referred to a gastroenterologist for the possibility that his symptoms were attributable to gastroesophageal reflux disease. An upper gastrointestinal endoscopic examination did not reveal evidence of esophagitis. The patient was begun on a histamine 2 blocker at bedtime, but this medication did not help his symptoms. After a third episode of chest pain necessitating admission to a coronary care unit, the thought arose that the patient's pains could be related to "esophageal spasm." Esophageal manometry revealed a low lower esophageal sphincter pressure at 8 mm Hg (normal 10 to 45 mm Hg). All contractions in the body of the esophagus were peristaltic, with normal amplitude and duration. An acid perfusion (Bernstein) test and Tensilon provocation did not bring on his chest pain. The patient next underwent prolonged esophageal pH monitoring done as an outpatient. Prior to this study he was encouraged to have normal activity, including eating large late night meals and possibly playing a game of racquetball. This study revealed minimally increased acid exposure times in both the upright (patient 9.2 percent, normal less than 8.1 percent) and supine positions (patient 6 percent, normal less than 3 percent). More important, the patient had six episodes of chest pain associated with belching from 7 to 11 P.M. on the evening of the study. Five of those episodes showed an excellent correlation between chest pain and acid reflux (pH less than 4.0) (Figure 11.1). After reviewing these test results, the patient was begun on a vigorous antireflux regimen, including omeprazole 20 mg BID. Over the last year, he has been essentially free of chest pain.

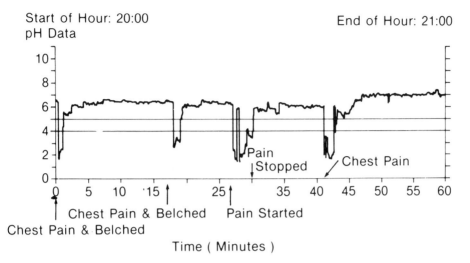

Start of Hour: 20:00
pH Data

End of Hour: 21:00

**Figure 11.1.** A 1 hour segment of R.J.'s ambulatory esophageal pH tracing showing the association of chest pain and belching with acid reflux episodes. Symptoms resolved with an antireflux regimen, including a histamine 2 blocker twice a day, after breakfast and 30 minutes after the evening meal.

# MECHANISMS FOR ACID-INDUCED CHEST PAIN

Reflux of gastric acid is usually associated with substernal burning discomfort, i.e., heartburn. However, chest pain may frequently accompany heartburn or be the only symptom of gastroesophageal reflux disease. The etiology of these complaints is not well understood. Reflux of bile salts and esophageal balloon distention may bring on some of these symptoms, suggesting that distention rather than acid alone may be causing these complaints. Other studies have suggested that esophageal motility abnormalities may be evoked by acid perfusion,[4] but more recent studies found that acid-induced pain usually is not associated with important changes in motility.[5] The acid sensitivity observed during intraesophageal acid perfusion or pH monitoring suggests that the major mechanism of pain is related to stimulation of mucosal chemoreceptors by the acid contents of the stomach. However, these receptors may not reside superficially, because topical anesthetics fail to alter the pain response.

Although acid reflux can cause heartburn and chest pain, the relationship between individual reflux episodes and these symptoms is poor. For example, postprandial reflux is common in healthy persons but symptoms are rare. Among esophagitis patients, intraesophageal pH monitoring frequently shows excessive periods of acid reflux, but patients usually experience heartburn or chest pain in less than 20 percent of reflux episodes. Moreover, one third of patients with Barrett's esophagus, the most extreme form of reflux disease, are acid-insensitive. Therefore, acid-induced symptoms may require more than just esophageal acid exposure. Mucosal disruption with inflammation may be contributory, but the majority of symptomatic patients have a normal esophagus by endoscopic examination and biopsy results. Hydrogen ion concentration could be pivotal in symptom production. A recent study found that all 25 reflux patients had reproduction of their symptoms during intraesophageal infusion of solutions of pH 1.0 and 1.5, but only half had pain with solutions of pH 2.5 to 6.0.[6] Other factors possibly influencing reflux-induced symptoms may include the acid clearance mechanism, salivary bicarbonate concentration, the total amount of acid exposure, the frequency of pain, and the interactions of pepsin with acid.

# INITIAL EVALUATION OF SUSPECTED ACID-RELATED CHEST PAIN

Evaluation of patients with recurrent chest pain should always begin with exclusion of significant cardiac disease. This is particularly important for several reasons. The most obvious is the potential life-threatening nature of coronary artery disease. In addition, the prevalence of both cardiac and gastroesophageal reflux disease increases as persons grow older. In one report, esophageal disease was found in as many as 50 percent of patients with coronary artery disease.[7] Thus, both problems may coexist and complicate the diagnostic evaluation. Furthermore, coexisting cardiac and esophageal disease may produce chest pain via complex interactions, thereby confusing the physician. For example, esophageal acid perfusion has been shown to produce ischemic electrocardiographic changes and increased myocardial workload in patients with coronary artery disease.[8] The extent of the cardiac work-up should be determined by the patient's age, family history, and risk factors for cardiac disease. Although heart disease in younger patients usually can be excluded by a normal electrocardiogram, an exercise stress test, and an echocardiogram, the coronary angiogram remains the gold standard.

## Clinical History

The clinical history frequently does not differentiate between cardiac and reflux-induced chest pain. For example, gastroesophageal reflux may be triggered by exercise,[9] and produce exertional chest pain mimicking angina pectoris, even during treadmill testing.[10] Symptoms suggesting an esophageal origin of chest pain include pain that continues for hours; retrosternal pain without lateral radiation; pain that interrupts sleep or is meal-related; pain that is relieved by antacid agents; or the presence of other esophageal complaints.[11] For example, in a recent study of 100 consecutive patients referred to our esophageal laboratory by cardiologists for noncardiac chest pain, we found that 74 percent had a history of heartburn, 67 percent had regurgitation, 49 percent had dysphagia, and 14 percent had odynophagia. However, 11 percent of the patients had no esophageal symptoms other than their chief complaint of chest pain.[12] Unfortunately, as many as 50 percent of patients with cardiac pain may have one or more symptoms of esophageal pain.[11]

## Upper Gastrointestinal Radiography and Endoscopy

The presence of erosive esophagitis and other structural lesions of the upper gastrointestinal tract possibly causing chest pain is best excluded by barium studies or endoscopy. If erosive or ulcerative esophagitis is found, acid reflux disease can be presumed to be causing the chest pain, and appropriate management can be begun. However, findings of mucosal hyperemia, increased vascularity, and friability are only "soft" signs of reflux disease, and further testing to define an acid-related cause of chest pain is required (*see* Case Study No. 2). As with other forms of reflux disease, our experience has been that the vast majority of patients with acid-related chest pain have a normal endoscopic examination. Therefore, we must turn to other tests to define the presence of abnormal acid reflux and its correlation with chest pain.

## Acid Perfusion (Bernstein) Test

Introduced in 1958 by Bernstein and Baker,[13] the acid perfusion test in attempts to identify esophageal acid exposure as a potential mechanism for pain. The test is done by placing a nasogastric tube midway down the esophagus and alternating infusions of saline with 0.1 N hydrochloric acid at a rate of 6 to 8 ml per minute. Saline should not cause chest pain and acid must replicate the patient's chest pain for findings to be called positive. Although the outcome of the test is frequently positive in patients with classic symptoms of heartburn, the prevalence of a positive acid perfusion test in patients with chest pain has varied considerably in the literature. For example, Behar and coworkers[14] found a prevalence of a positive acid perfusion test to be 100 percent in 11 patients with chronic chest pain, whereas Katz and colleagues[15] observed that only 61 out of 910 patients (6.7 percent) referred to an esophageal laboratory had a positive acid perfusion test replicating their chest pain complaints. Recent studies comparing the acid perfusion test and prolonged pH monitoring suggest that the former study may be a specific although insensitive test. We have prospectively compared these two tests in 75 consecutive noncardiac chest pain patients who had both an acid perfusion test and chest pain during prolonged pH monitoring.[16] Although it had excellent specificity (90 percent), the acid perfusion test had poor sensitivity (36 percent) for replicating the patient's acid-induced chest pain. Thus, a positive test result reliably indicates that the patient's chest pain is the result of esophageal acid sensitivity. However, a negative test result does not exclude acid-sensitive chest pain, and the patient should be considered for prolonged pH monitoring.

# ROLE OF PROLONGED 24-HOUR ESOPHAGEAL pH TESTING

The development of small, compact ambulatory esophageal pH monitors now allows us to study our patients in the home or workplace. This technology permits pH studies to be performed in more natural settings whereby patients can eat their regular diets, smoke, drink alcohol, and exercise—all or any of which may be important factors in the precipitation of their chest pain.

## Prevalence of Acid-Related Chest Pain

Recent studies with prolonged esophageal pH monitoring suggest that gastroesophageal reflux disease is the most common esophageal cause of noncardiac chest pain (Table 11.1). In 1982, DeMeester and coworkers[17] noted abnormal gastroesophageal reflux dur-

TABLE 11.1. AMBULATORY ESOPHAGEAL pH STUDIES OF PATIENTS WITH
NONCARDIAC CHEST PAIN

| Investigators | No. of Patients | Abnormal Gastroesophageal Reflux | Pain with Gastroesophageal Reflux |
|---|---|---|---|
| DeMeester et al.[17] | 50 | 23 (46%) | 12 (24%) |
| DeCaestecker et al.[18] | 50 | 14 (28%) | 6 (12%) |
| Janssen et al.[19] | 60 | 13 (22%) | 13 (22%) |
| Peters et al.[20] | 24 | 8 (33%) | 10 (42%) |
| Schofield et al.[21] | 52 | 11 (21%) | 9 (17%) |
| Richter et al.[16] | 45 | 20 (44%) | 14 (31%) |
| Soffer et al.[22] | 20 | ND* | 9 (45%) |
| Hewson et al.[12] | 100 | 48 (48%) | 50 (50%) |
| Ghillebert et al.[23] | 50 | ND* | 15 (30%) |
| Nevens et al.[24] | 37 | 15 (41%) | 5 (14%) |
| Paterson et al.[25] | 25 | ND* | 10 (40%) |
| Totals | 513 | 152 (37%) | 153 (30%) |

*ND = Not Done

ing 24 hour pH monitoring in 23 (46 percent) of 50 patients who had angina-type chest pain and normal coronary angiograms. Of the 13 patients with reflux who had chest pain during the study, the onset of pain in 12 coincided with the reflux episode on the pH record. At the time, this seemed like an inordinately high frequency of reflux for this kind of patient, but further studies have provided support for this observation. De Caestecker and colleagues[18] found an abnormal amount of reflux in 14 (28 percent) of 50 patients who had unexplained chest pain. Other reports have shown that in 5 to 50 percent of patients with normal coronary anatomy, their typical chest pain occurs during reflux episodes.[12,16,19-25] We have recently compared the diagnostic capabilities of traditional esophageal tests with 24 hour pH monitoring in 100 consecutive patients referred by cardiologists to the esophageal laboratory for evaluation of noncardiac chest pain.[12] Patient R.J. in Case No. 1 was one of these patients. Esophageal manometry and acid perfusion and edrophonium tests identified the esophagus as definitely contributing to chest pain in 28 percent of patients. By comparison, 24 hour pH testing was significantly superior ($p < 0.01$) to traditional esophageal tests by defining 50 percent of patients as having an esophageal cause of their chest pain (Figure 11.2). In the 83 patients with spontaneous chest pain during 24 hour pH testing, 37 patients (46 percent) had abnormal reflux parameters and 50 patients (60 percent) had a positive correlation between chest pain and acid reflux episodes (mean positive score 56 percent, range 6 to 100 percent). Additionally, 11 patients denied having chest pain during this study but had abnormal reflux parameters suggesting a possible esophageal cause for their symptoms. More than half of the patients with a positive symptom index had normal amounts of acid exposure during prolonged pH testing. Thus, these patients may be hypersensitive to a physiologic amount of acid reflux, although abnormal quantities of acid reflux could have been missed by the pH probe (Figure 11.3).

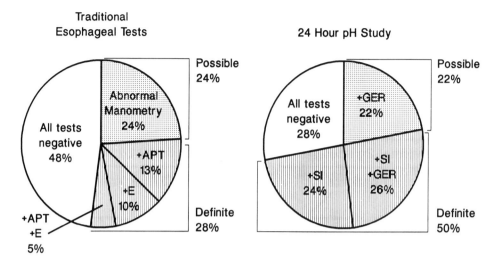

**Figure 11.2.** Comparison between the diagnostic yield of traditional esophageal tests (manometry and provocative testing with acid and edrophonium) and 24 hour esophageal pH monitoring in 100 consecutive patients with noncardiac chest pain. An esophageal test was defined as "definitely" identifying the esophagus as a cause of chest pain only if provocative testing replicated the patient's identical chest pain or the patient's spontaneous chest pain correlated with at least one episode of acid reflux (positive symptom index) during 24 hour pH monitoring. Otherwise, abnormal manometry or acid reflux parameters without associated chest pain only suggested the esophagus as a "possible" cause of chest pain. Esophageal pH monitoring with symptom index was significantly superior ($p < 0.001$; McNemar's test) to traditional esophageal tests in definitely identifying the esophagus as contributing to these patients' chest pain. Abbreviations: + APT = positive acid perfusion test, + E = positive edrophonium test, + GER = abnormal acid reflux parameters, + SI = positive symptom index.

## Exertional Chest Pain and Acid Reflux

Exertional chest pain is a classic symptom of coronary artery disease and an important differential finding. However, recent studies[9,10] suggest that exertion may exacerbate gastroesophageal reflux disease in reflux patients and induce acid reflux in healthy subjects. Schofield and colleagues[10] used prolonged esophageal pH monitoring during exercise treadmill testing in evaluating 52 patients with chest pain and normal coronary angiograms to assess the frequency of reflux-associated pain. Eleven patients had abnormal baseline reflux parameters, 9 of whom (82 percent) had their typical chest pain occur coincident with reflux during treadmill testing. More interesting, 13 patients with normal baseline reflux parameters experienced exertional chest pain while on the treadmill that occurred during acid reflux episodes. None of these patients showed electrocardiographic evidence of ischemia. Overall, 44 percent of patients studied were shown to have their chest pain reproduced simultaneously with acid reflux.

The study by Schofield and coworkers[10] emphasizes the importance of investigating

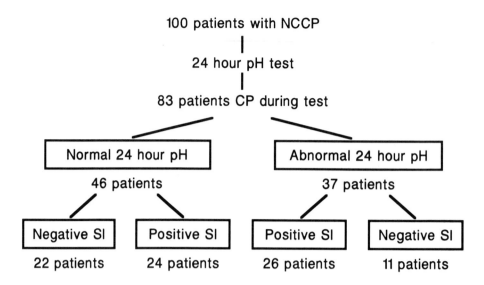

**Figure 11.3.** Results of ambulatory 24 hour esophageal pH monitoring in 100 consecutive patients referred by cardiologists for the evaluation of noncardiac chest pain. Seventeen patients did not experience chest pain during the pH test. Of the remaining 83 patients, 55 percent had normal acid reflux parameters, whereas 45 percent had abnormal reflux parameters. Over twice as many patients with abnormal reflux parameters had at least one of their chest pain episodes coincident with acid reflux, i.e., a positive symptom index (SI). Despite only physiologic amounts of acid reflux present in the other group, 24 (52 percent) patients had a positive symptom index. Overall, in 50 patients at least one episode of their chest pain was closely associated with gastroesophageal reflux.

patients during the physiologic episodes that bring on their chest pain. However, we do not believe that treadmill testing and pH monitoring need to be routinely done in all patients with suspected acid-related chest pain. Rather, the medical personnel performing the pH study must take a careful history to identify precipitators of chest pain and encourage patients to carry out these same activities during the monitoring period. Using this approach, nearly 85 percent of patients undergoing pH monitoring for the evaluation of noncardiac chest pain in our laboratory experienced the replication of their symptoms during outpatient studies. Treadmill testing should be reserved for those patients with exercise-induced symptoms who do not have pain during the standard outpatient pH study.

## Even a Negative Study May Be Helpful

### Case Study No. 2

W.J. is a 44 year old male who came to his family physician with a 1 month history of substernal chest pain radiating into his arms and shoulders associated with abdominal swelling and flatus. His symptoms occurred primarily upon exertion. Most vividly, he recalled several times that his pain came on while walking his dog up and down several hills near his home. The patient also noted similar, although not as strong, pains occasionally after meals and when he was bending over working in the garden. The patient's only esophageal complaint was occasional heartburn, which came on after heavy meals when he had had too much to eat or smoke. Physical examination was unremarkable except for moderate obesity.

Evaluation by the patient's family physician included a normal resting electrocardiogram and an exercise stress test done through 3 minutes of stage IV on the standard Bruce protocol. Upper gastrointestinal endoscopy revealed a large hiatas hernia with distal esophageal edema and hyperemic streaks. The patient was placed on an antireflux regimen, which included elevation of the head of the bed, no heavy meals 3 hours prior to bedtime, a histamine 2 blocker twice a day, and metoclopramide, 10 mg, before meals. Additionally, the patient was cautioned to avoid heavy exercising in the evening. This regimen resulted in a 50 to 75 percent improvement of W.J.'s symptoms. However, he did not like being on medication and questioned whether surgery might not be the best approach for his acid reflux disease.

The patient was referred to a gastroenterologist for further evaluation prior to surgery. Baseline esophageal manometry revealed a normal lower esophageal sphincter pressure at 20 mm Hg, with peristaltic contractions in the body of the esophagus. Twenty-four hour esophageal pH testing was performed on an outpatient basis, with the subject being encouraged to perform all the normal activities that had brought on his symptoms in the past. W.J. went home, ate several heavy meals, and also walked his dog up and down the hills around his home. During the study, the patient was noted to have a physiologic amount of acid reflux after meals (patient 5.0 percent, normal less than 8.1 percent) with no evidence of acid reflux at night. However, the patient had four episodes of chest pain that were not associated with acid reflux. Two of these episodes were brought on while walking his dog and interestingly were also associated with complaints of heartburn (Figure 11.4). On further questioning, the patient admitted that the improvement of his symptoms more closely correlated with decreasing his daily exercise than with his antireflux therapy. Given this additional history and a lack of correlation of chest pain with acid reflux, the patient was referred to a cardiologist for possible cardiac catheterization. Subsequent coronary angiography revealed coronary artery disease in three vessels. The patient is symptom-free after having undergone coronary artery bypass surgery.

Although it is relatively dramatic, this case emphasizes that a negative pH study result may be helpful if the patient's classic chest pain is reproduced in an outpatient setting.

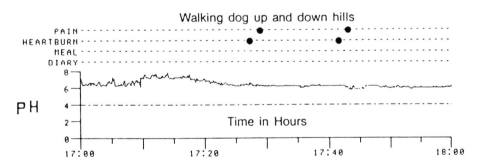

**Figure 11.4.** A 1 hour segment of W.J.'s ambulatory esophageal pH study while he was walking his dog up and down hills near his home. Although the patient had chest pain and heartburn, no acid reflux was recorded. Subsequent cardiac catheterization confirmed that severe coronary artery disease was the cause of these symptoms.

Therefore, we must identify and encourage our patients to recreate the activities that have been associated with their chest pain. This should be done even if it means ingesting prohibited acidic foods, eating meals late at night or snacking, or performing activities that have been restricted in the past. This case also underscores the caution that must be observed in interpreting minor esophageal mucosal changes observed during endoscopic examination.

## Gastroesophageal Reflux Disease and Heart Disease

### Case Study No. 3

J.P. is a 44 year old man with a 2 year history of coronary artery disease. The diagnosis was initially made in January 1987, after a 1 year history of intermittent substernal chest pain both at rest and during exertion. One severe episode associated with left arm numbness, shortness of breath, and nausea resulted in admission to a coronary care unit. The admission cardiogram showed ST segment depression in the inferior leads, which resolved after the patient's pain improved on ingesting three sublingual nitroglycerin tablets. Cardiac enzyme levels were normal. Because of the crescendo quality of the angina, cardiac catheterization was performed. This study revealed 70 percent narrowing of the circumflex artery and 50 percent narrowing of the distal left anterior descending artery, with a normal ejection fraction of 78 percent. The patient's cardiologist chose to begin medical therapy with nitroglycerin paste 1 inch every 6 hours and sublingual nitroglycerin.

The patient's past medical and social histories were remarkable for a 60 pack year history of tobacco and a positive family history for early myocardial infarction. His only other medical complaint was occasional postprandial heartburn.

J.P. did well for about 1 year, until a "different" chest pain began primarily after meals and at night. This pain had a burning quality but was distinct from his heartburn. It was substernal in location, did not radiate, and seemed to catch his breath and cause him to choke and cough. Initially, his cardiologist thought his heart disease had worsened and added nifedipine, 10 mg by mouth three times a day, to his antianginal medical regime. When his pains worsened, a second cardiac catheterization was performed, but this angiogram was essentially unchanged from 1 year before.

J.P. was referred to a gastroenterologist for further evaluation of his atypical chest pain. Endoscopy revealed a hiatas hernia but no evidence of esophagitis. Because of the suspicion that his chest pain might be multifactorial in origin, combined esophageal pH and Holter monitorings were performed on an outpatient basis. The pH study revealed normal acid exposure during the day but increased acid exposure (patient 10 percent; normal less than 3 percent) at night. Three episodes of burning, smothering chest pain occurred during the study—two while the patient was sawing wood after a light dinner and one that awoke him about 1 A.M. at night. Figure 11.5 illustrates the simultaneous pH and electrocardiogram tracings during two of these episodes of chest pain. The study was interpreted as initially showing ischemic chest pain brought about by the exercise of sawing wood. The first ischemic episode was followed 2 minutes later by an episode of acid reflux, which was probably caused by the patient's bending over while exercising soon after a meal. The episode at night was a similar pain that was associated with a long episode of supine reflux (30 minutes) but no ischemic ECh changes.

These studies confirmed that both gastroesophageal reflux and coronary artery disease were contributing to J.P.'s chest pain complaints. The patient was encouraged to elevate the head of his bed and not to eat for several hours before retiring or to exercise immediately after meals. A proton pump inhibitor was added to his medical regimen in the A.M. Unfortunately, neither the patient nor his physician could adequately define which chest pains were cardiac versus reflux disease in origin. Nevertheless, his smothering chest pain improved and no longer was a problem at night.

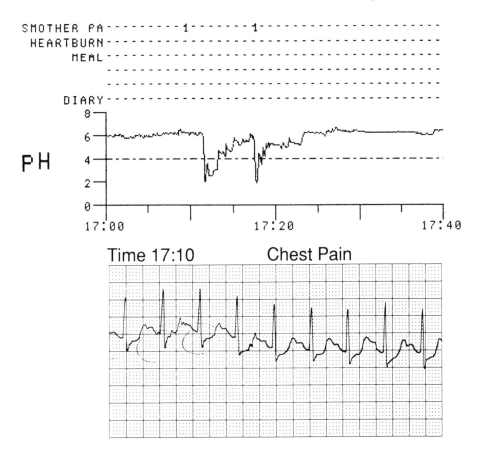

**Figure 11.5.** A 40 minute segment of J.P.'s ambulatory esophageal pH and Holter study obtained while he was sawing wood after a light dinner. Smothering chest pain began at 17:10 associated with S-T segment depression on the electrocardiogram strip. Approximately 2 minutes later the pH tracing recorded an episode of acid reflux, probably precipitated by the patient's bending over while exercising soon after a meal. The sequence was repeated again when J.P. resumed his exercise. Although one might postulate that angina could have precipitated the reflux episode, more likely the common denominator was exercise on a full stomach coincidentally bringing on the two problems.

Gastroesophageal reflux disease and heart disease both increase in prevalence as people grow older.[7] In addition, drugs used to treat coronary artery disease, such as nitrates, beta blockers, and calcium channel blockers, may decrease lower esophageal sphincter pressure and contraction amplitudes, thereby aggravating or predisposing to gastroesophageal reflux. Therefore, it should not be surprising that reflux-induced chest pain can also occur in patients with obstructive coronary artery disease. Since these pain syndromes can be identical, objective data are essential to determine the source. Prolonged ambulatory esophageal pH monitoring and simultaneous multilead cardiac monitoring (Holter) permit the concurrent recording of both esophageal and cardiac events. In a group of 36 patients with angiographically documented obstructive coronary artery disease, we found that 16 patients (44 percent) had abnormal acid reflux parameters, and 16 of 32 patients (50 percent) experienced chest pain episodes coinciding with acid reflux during prolonged monitoring. Four patients also had evidence of ischemia, two with associated acid-induced chest pain and one with abnormal reflux parameters. Therapy based on these findings resulted in coronary artery bypass grafting in two patients, increased nitrates in one patient, and vigorous antireflux therapy in 14 patients. In the last group, 10 of 14 patients with reflux-associated pain improved while receiving famotidine, 40 mg twice a day, or omeprazole, 20 mg every morning.[26]

# MANAGEMENT OF GASTROESOPHAGEAL REFLUX RELATED CHEST PAIN

Although the history and pH monitoring may suggest acid reflux as the cause of chest pain, aggressive antireflux therapy is the only way to confirm the diagnosis. It is important to emphasize, however, that until recently there has been no randomized, controlled treatment trials specifically examining patients with GERD presenting with the primary complaint of chest pain.

Several uncontrolled reports suggest that high dose $H_2$ blockers or omeprazole are effective treatments. DeMeester and associates[17] noted that 7 of 12 patients with chest pain and gastroesophageal reflux improved on medical therapy, including antacids and cimetidine. In a second study, 13 patients with chest pain and gastroesophageal reflux disease were treated in an open study with high dose ranitidine (150 mg three times per day to 300 mg four times per day).[27] The mean symptom score improved from 2.87 to 0.68 after eight weeks of therapy, and seven patients had complete resolution of their chest pain. In the third study, 13 of 18 patients (75 percent) with coronary artery disease and refractory chest pain thought to be related to acid reflux had marked improvement after vigorous acid suppression with high dose $H_2$ blockers or omeprazole for eight weeks.[26]

Two uncontrolled series suggest that Nissen fundoplication surgery provides relief of chest pain in patients with gastroesophageal reflux disease. DeMeester et al.[17] noted improvement of chest pain in 10 of 11 patients following antireflux surgery. Similarly, Bancewicz and coworkers[28] reported that 23 of 24 patients with gastroesophageal reflux disease and chest pain had resolution of their pain after Nissen fundoplication with a follow-up period of 3 to 6 years.

Achem and coworkers[29] performed a double-blind, placebo controlled, parallel study in 34 consecutive patients with chest pain and gastroesophageal reflux defined by 24 hour pH monitoring. Subjects were randomized to omeprazole, 20 mg two times a day or placebo. After 8 weeks of treatment, patients on omeprazole obtained significantly more improvement in the fraction of chest pain time and severity when compared to placebo. Analysis of patient outcome found that 13 (81percent) of the subjects in the omeprazole reported overall symptomatic improvement versus 1 (6 percent) in the placebo group.

The author's approach to the patient with chest pain and gastroesophageal reflux disease is as follows: After 24 hour pH monitoring suggests the diagnosis, patients are aggressively treated with lifestyle changes and a proton pump inhibitor. My choice is omeprazole 20 mg b.i.d. or lansoprazole 30 mg b.i.d. used for one to two months. If symptoms markedly improve or resolve, the diagnosis is confirmed and the patient titrated down to the lowest dose of required medication, either proton pump inhibitor, histamine 2 blocker or promotility drug. If symptoms are not improved, repeat 24 hour pH monitoring is performed to ascertain whether or not adequate acid suppression is on board. If this is not the case, then the dose of medication is doubled. In appropriate patients, antireflux surgery may be considered but only for those individuals in whom aggressive medical therapy has completely relieved their chest pain.

# SUMMARY

Patients with recurrent chest pain who are free of significant coronary artery disease account for 10 to 30 percent of patients undergoing coronary angiography. Recent studies suggest that gastroesophageal reflux disease may be very common in these patients. Unfortunately, a careful history does not distinguish chest pain arising from a cardiac versus an esophageal source. Therefore, all patients must undergo a thorough cardiac evaluation before assuming that acid reflux is the cause of their complaints. Initial gastroenterology evaluation usually includes upper gastrointestinal endoscopy and/or barium studies, possibly with acid perfusion (Bernstein) testing. However, the more sensitive and specific test for acid reflux disease is prolonged esophageal pH monitoring. This study quantifies the amount of acid reflux but more importantly identifies the relationship between chest pain and acid reflux episodes. Patients should be studied in the outpatient setting, with emphasis placed on performing activities that replicate their chest pain. In this situation, even a negative test result may be helpful. The addition of Holter monitoring may be useful in patients with coronary artery disease and persistent chest pain, in whom acid reflux may be another factor accounting for their intractable symptoms.

# REFERENCES

1. Ockene IS, Shay MJ, Alperts JS, et al: Unexplained chest pain in patients with normal coronary arteriograms. A follow-up study of functional status. *N Engl J Med* 30:1249–1252, 1980.
2. Lawrence L: National Center for Health Statistics. Detailed diagnoses and procedures for patients discharged from short-stay hospitals in the United States. 1984 Vital Health Statistics Series 13, No. 86, 1986.
3. Richter JE, Bradley LA, Castell DO: Esophageal chest pain: Current controversies in pathogenesis, diagnosis and therapy. *Ann Int Med* 110:66–78, 1989.
4. Siegel CI, Hendrix TR: Esophageal motor abnormalities induced by acid perfusion in patients with heartburn. *J Clin Invest* 42:686–695, 1963.
5. Richter JE, Johns DN, Wu WC, et al: Are esophageal motility abnormalities produced during intraesophageal acid perfusion test? *JAMA* 253:1914–1917, 1985.
6. Smith JL, Lankai E, Opekum AR, et al: Sensitivity of the esophageal mucosa to pH in gastroesophageal reflux disease. *Gastroenterology* 96:683–689, 1989.
7. Svensson O, Stenport G, Tibbling L, et al: Oesophageal function and coronary angiogram in patients with disabling chest pain. *Acta Med Scand* 204:173–178, 1978.
8. Mellow MH, Simpson AG, Watt L, et al: Esophageal acid perfusion in coronary artery disease: Induction of myocardial ischemia. *Gastroenterology* 83:306–312, 1983.

9. Clark CS, Kraus BB, Sinclair J, et al: Gastroesophageal reflux induced by exercise in healthy volunteers. *JAMA* 261:3599–3601, 1989.

10. Schofield PM, Bennett DH, Whorwell PJ, et al: Exertional gastroesophageal reflux: A mechanism for symptoms in patients with angina pectoris and normal coronary angiograms. *Br Med J* 294:1459–1461, 1987.

11. Alban-Davies H, Jones DB, Rhoades J, et al: Angina-like esophageal pain: Differentiation from cardiac pain by history. *J Clin Gastroenterol* 7:477–481, 1985.

12. Hewson EG, Sinclair JW, Dalton CB, Richter JE. 24-hour esophageal pH monitoring. The most useful test for evaluating non-cardiac chest pain. *Am J Med* 90:576–83, 1991.

13. Bernstein LM, Baker LA: A clinical test for esophagitis. *Gastroenterology* 34:760–781, 1958.

14. Behar J, Biancini P, Sheahan DG: Evaluation of esophageal tests in the diagnosis of reflux esophagitis. *Gastroenterology* 71:9–15, 1976.

15. Katz PO, Dalton CB, Richter JE, et al: Esophageal testing in patients with noncardiac chest pain and/or dysphagia. *Ann Int Med* 106:593–597, 1987.

16. Richter JE, Hewson EG, Sinclair JW, et al: Acid perfusion test and 24-hr esophageal pH monitoring with symptom index: Comparison of tests for esophageal acid sensitivity. *Dig Dis Sci* 1991; 36:565–71.

17. DeMeester TR, O'Sullivan GC, Bermudez G, et al: Esophageal function in patients with angina-type chest pain and normal coronary angiogram. *Ann Surg* 196:488–498, 1982.

18. DeCaestecker JS, Brown J, Blackwell JN, et al: The oesophagus as a cause of recurrent chest pain: Which patients should be investigated and which tests should be used? *Lancet* 2:1143–1146, 1985.

19. Janssen J, Vantrappen G, Ghillebert G: 24-Hour recording of esophageal pressure and pH in patients with non-cardiac chest pain. *Gastroenterology* 90:1978–1984, 1986.

20. Peters L, Maas L, Petty D, et al: Spontaneous non-cardiac chest pain: Evaluation by 24-hour ambulatory esophageal motility and pH monitoring. *Gastroenterology* 94:878–886, 1988.

21. Schofield PM, Whorwell PJ, Brooks NH, et al: Oesophageal function in patients with angina pectoris: A comparison of patients with normal coronary angiogram and patients with coronary artery disease. *Digestion* 42:70–78, 1989.

22. Soffer EE, Scalabrini P, Wingate DL: Spontaneous noncardiac chest pain: Value of ambulatory esophageal pH and motility monitoring. *Dig Dis Sci* 96:683–89, 1989.

23. Ghillebert G, Janssen J, Vantrappen G, et al: Ambulatory 24 hour esophageal pH and pressure readings vs. provocative tests in the diagnosis of chest pain of oesophageal origin. *Gut* 31:728–735, 1990.

24. Nevens F, Janssen J, Piessens J, et al: Prospective study on prevalence of esophageal chest pain in patients referred on an elective basis to a cardiac unit for suspected myocardial ischemia. *Dig Dis Sci* 36:229–235, 1991.

25. Paterson WG, Abdollah T, Beck IT, et al: Ambulatory esophageal manometry, pH-metry and Holter monitoring in patients with atypical chest pain *Dig Dis Sci* 38:795–801, 1993.

26. Singh S, Richter JE, Hewson EG, et al: The contribution of gastroesophageal reflux to chest pain in patients with coronary artery disease. *Ann Intern Med* 117:824–830, 1992.

27. Stahl WA, Beton RR, Johnson CS, et al: High-dose ranitidine in the treatment of patients with non-cardiac chest pain and evidence of gastroesophageal reflux. *Gastroenterology* 102:162, 1992.

28. Bancewicz J, Osugi H, Marples M: Clinical implications of abnormal oesophageal motility. *Br J Surg* 74:416–420, 1987.

29. Achem SR, Koltz BE, Richter JE, et al: Treatment of acid related non-cardiac chest pain: A double-blind placebo controlled study of omeprazole vs. placebo. *Gastroenterology* 104:51, 1993.

# 12

# Otolaryngologic Manifestations of GERD

Peter J. Kahrilas, M.D.
Gulchin A. Ergun, M.D.

The linkage between gastric reflux and chronic otolaryngological disorders is a relatively recent concept. The German pathologist, Rudolf Virchow, attributed contact ulcers and granulomata over the vocal processes of the arytenoid cartilages to voice abuse and named it "pachydermia verrucosa laryngis," a term now used interchangeably with "contact ulcer" or "contact ulcer granuloma."[1] Chevalier Jackson noted exposed cartilage at the base of contact ulcers and elaborated on the "hammer and anvil" theory of pathogenesis whereby mucosal injury and granuloma formation occur at the point where the vocal processes impact on each other during speech.[2] It was not until 1968 that Jerrie Cherry first implicated irritation from the esophagopharyngeal reflux of gastric juice as the key predisposing factor leading to irritation or injury of the pharyngeal or laryngeal mucosa.[3] Since then, numerous clinical observations have supported this contention. Pathophysiologic data suggest that the propensity for refluxate to reach the proximal esophagus with consequent regurgitation into the hypopharynx[4] discriminates reflux patients with and without otolaryngologic manifestation of GERD. The spectrum of associated pathologic conditions that are implicated includes reflux laryngitis, subglottic stenosis, and laryngeal cancer.

Substantial evidence suggests a relationship between acid reflux and laryngitis evident by interarytenoid erythema, plaques, nodularity, and contact ulcer (Figure 12.1). These patients complain of chronic dysphonia, cough, globus sensation, frequent throat clearing, or sore throat without identifiable allergic or infectious causes. In a large series, laryngoscopic lesions, ranging from edema to granulomas, were limited to the posterior third of the vocal cords and interarytenoid area. Histologically, the laryngeal epithelium exhibited hyperplasia of the prickle cell or basal cell layer with abundant lymphocytic and plasma-cell infiltration.[5] Many case reports and two recent uncontrolled therapeutic trials have reported recovery from chronic laryngitis following antireflux therapy.[6,7] However, despite the association between posterior laryngitis and reflux disease, overt esophagitis is paradoxically absent in most of these individuals.[8]

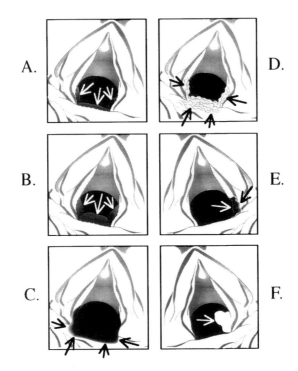

**Figure 12.1.** Line drawings of reflux laryngitis pathology. The 6 images demonstrate mild posterior glottic erythema (A), severe posterior glottic edema (B), severe post glottic erythema (C), severe posterior glottic nodularity (D), contact ulceration (E), and posterior glottic granuloma (F). In each case the luminal arrows (white) or the black arrows overlying the posterior glottis highlight the abnormality. Modified from: Shaw GY, Searl JP. Laryngeal manifestations of gastroesophageal reflux before and after treatment with omeprazole. *South Med J* 1996; In Press, with permission.[15]

Tracheal stenosis in adults most commonly occurs as a complication of endotracheal intubation. The pathogenesis begins with mucosal injury, which is followed by an inflammatory response and a hyperplastic reparative process.[9] The vocal process of the arytenoids and the posterior cricoid are the sites most often injured as a result of intubation. Several animal studies demonstrate the development of subglottic stenosis when disruption of the mucosa is followed by periodic exposures of the exposed cartilage to gastric secretions.[8] Patients intubated beyond 48–72 hours are most prone to develop laryngeal mucosal injuries, and esophageal pH recordings suggest that pharyngeal reflux of gastric contents is common in recumbent intubated patients,[10] perhaps explaining the incidence of subglottic stenosis in such patients.

The worst potential risk of reflux laryngitis is the development of squamous cell cancer of the larynx. The overwhelming majority of laryngeal cancer cases can be linked to cigarette smoking with excessive alcohol consumption being either an independent or a potentiating factor.[11] Recently, it has been suggested that reflux laryngitis also plays a role in the pathogenesis of some cases of squamous cell cancer of the larynx. In 1983 Olsen reported five cases of vocal process squamous cell cancer that resembled contact ulcer with granuloma; all five patients were smokers.[12] Subsequently two groups of investigators reported a series of about 20 patients each who were lifetime nonsmokers but had chronic reflux and developed laryngeal squamous cell cancer.[13,14] In the Ward series, three patients with chronic reflux laryngitis were serially observed as their lesions evolved from benign laryngeal lesions to laryngeal cancer over periods of 5, 5, and 8 years.

# THE PATIENT HISTORY

Because there are no associated pathognomonic findings, the etiology of chronic laryngeal or pharyngeal symptomatology due to reflux is to some degree pursued as a diagnosis of exclusion. Both the chronicity of the problem and a careful history are used to exclude potential allergic or infectious etiologies. Thereafter, the patient with persistent chronic symptoms becomes a candidate for the diagnosis. The primary symptoms of GERD-associated otorhinolaryngologic disorders are chronic or intermittent hoarseness, chronic throat clearing, difficulty or discomfort in the throat on swallowing (cervical dysphagia), and the sensation of a foreign body or lump in the throat (globus sensation). In addition, patients may complain of persistent or recurring sore throat in the absence of an apparent infection, the sensation of "postnasal drip" with throat clearing, or cough in the absence of pulmonary or tracheobronchial disease. Patients commonly experience several of these symptoms, although it is usually one in particular that concerns them enough to prompt evaluation. Interestingly enough, the associated esophageal symptomatology, predominantly heartburn and/or regurgitation, is often minimal and not of major concern to the patient. The series recently reported by Shaw typifies the symptom mix experienced by this patient group (Table 12.1).[15]

**TABLE 12.1. SYMPTOM PROFILE AND RESPONSE TO THERAPY OF 96 PATIENTS WITH OTOLARYNGOLOGIC MANIFESTATIONS OF REFLUX DISEASE.[15]**

| Symptom | At Presentation % (severity 0–3 scale) | After Antireflux Therapy % (severity 0–3 scale) |
|---|---|---|
| Throat clearing | 88% (2.39) | 44% (1.48) |
| Globus sensation | 83% (2.37) | 36% (1.48) |
| Hoarseness | 60% (1.85) | 34% (1.09) |
| Chronic cough | 55% (1.60) | 28% (1.17) |
| Bad taste | 48% (1.59) | 20% (0.88) |
| Heartburn | 39% (1.07) | 10% (0.75) |
| Dysphagia | 31% (0.75) | 7% (0.33) |
| Odynophagia | 3% (0.24) | 0% (0.09) |

*Patients were treated for 12 weeks with omeprazole 20 mg bid. In each category the percentage of patients acknowledging the symptom as well as the mean symptom score of the entire group is indicated. Note that all symptoms significantly improved with therapy.

Although it is difficult to profile a typical patient with otolaryngologic manifestations of reflux disease, there are some recurring themes. Professional voice users, be they singers, teachers, or politicians all stress their larynges by intense voice usage and are also extraordinarily aware of minor decrements in voice function. Even minor dysfunction is likely to prompt them to seek medical attention because of the associated career impact. These individuals may or may not also be prone to the single most significant lifestyle factor that exacerbates this condition; late night eating and drinking. Another major patient group afflicted with reflux laryngitis, young professionals, will often develop reflux otolaryngologic symptomatology simply on the basis of this lifestyle pattern. These individuals work late, eat a late dinner, and then go to bed shortly thereafter. Their voice function is worst in the morning and improves somewhat during the course of the day. Again, esophageal symptomatology is generally minimal and these individuals derive significant benefit simply from lifestyle modification. A final point to make regarding symptoms of reflux laryngitis is that the symptom of neck pain lateral to the thyroid cartilage is relatively specific for contact ulcer of the vocal fold process on the arytenoid cartilage and, in association with other suggestive symptoms, is a very useful historical clue for the physician.

# PATIENT EVALUATION OTHER THAN pH MONITORING

## Laryngoscopy

The diagnosis of chronic laryngitis is based on a laryngeal examination by indirect examination utilizing either rigid or flexible instruments or by direct inspection under general anesthesia. Diffuse laryngitis may be either a simple hyperemia or a hyperplastic response of the laryngeal mucosa. Keratosis refers to the clinical description of white lesions characterized by an excess of keratin; when these lesions occur on the posterior part of the cords they constitute "classical posterior laryngitis." Two other inflammatory lesions implicated in the spectrum of "acid" laryngitis are the "contact ulcer" which occurs over the vocal processes on the medial edge of the vocal cord and vocal cord granuloma whose typical polypoid appearance is said to be pathognomonic of a response to trauma. Figure 12.1 illustrates the spectrum of laryngeal injury associated with reflux laryngitis.

The histologic findings associated with laryngoscopic abnormalities of "posterior chronic acid laryngitis" were detailed in an extensive series of 44 patients.[5] Common findings were swollen, reddish, and hypertrophic mucosa of the interarytenoid area and granulomas of various size with or without ulcers on one or both vocal processes. Histology of the lesions showed simple hyperplasia of the squamous epithelium, mostly due to hyperplasia of the prickle cell with inflammatory cell infiltration, rarely involving the basal cell layer. Keratinization was noted only occasionally. The sensitivity or specificity of these findings for acid reflux is not known because there has been no similar analysis of controls or of patients with other etiologies of laryngitis. Hence, *no single* laryngeal pathology has been unequivocally linked with reflux.

Two clinical studies to date have aimed to objectively describe the laryngoscopic signs of laryngitis utilizing a standardized scale and to ascertain whether or not the lesions responded to antireflux therapy. The more comprehensive series involved 68 patients whose laryngoscopic images were graded for the severity of erythema, edema, nodularity, ulceration and granuloma (Figure 12.1).[15] All patients were treated uniformly with omeprazole 20 mg b.i.d. for 12 weeks and the laryngoscopic evaluation repeated after therapy. Posterior glottic edema, nodularity, and erythema were the most severe and frequent findings; subjects with these findings experienced significant improvement post-therapy. Contact ulceration and granulomata were found in only three patients, highlighting the important point that most laryngoscopic signs of posterior laryngitis are subtle. These conclusions were consistent with the results from an earlier, smaller series using a slightly different grading scheme for laryngoscopy.[6] In that series, laryngoscopic findings also tended to be mild, with the observation that the more chronic and severe the laryngeal injury, the less likely a complete response to therapy. Taken together, these laryngoscopic studies suggest that, despite symptomatic severity, posterior laryngitis occurs in a continuum of severity which only uncommonly culminates in the formation of a contact ulcer or granuloma.

## Esophagoscopy

The utility of esophageal endoscopy in the evaluation of a patient with suspected reflux laryngitis is limited by its low sensitivity. Even in patients with esophageal symptoms of reflux disease, the sensitivity of esophagoscopy is less than 50 percent. For reflux laryngitis, the sensitivity is probably closer to 15 percent. Furthermore, esophagitis is often transient and can be conspicuously absent in severely symptomatic patients. On the other hand, esophageal mucosal erosions are essentially 100 percent specific for reflux disease. Thus, the detection of esophagitis greatly aids in patient management but a negative endoscopy does not. Thus, it should not be surprising that in treatment trials of posterior laryngitis, the majority of patients (73–81%) had endoscopically normal esophagi at entry to the studies.[6,16]

## Upper GI x-ray

While several clinical tests have been advocated over the years to evaluate gastroesophageal reflux disease, the barium swallow has fallen out of favor because of poor sensitivity in diagnosing any but the most severe grades of esophagitis (25–35%). A barium swallow probably has no role in evaluating the patient with suspected reflux laryngitis.

## Esophageal Manometry

Esophageal manometry is used to evaluate very limited aspects of esophageal function in a clinical setting; specifically, the integrity of primary peristalsis and lower esophageal sphincter (LES) function. It follows that the potential yield of the test is limited to a few pathological aberrations such as absent or weak peristalsis, disordered peristalsis, abnormalities of LES tone, or impaired LES relaxation. Whereas one or more manometric aberrations may occur in gastroesophageal reflux disease (impaired peristalsis, LES hypotension, decreased LES length, or an increased number of transient LES relaxations),[17] the real question from a clinical perspective is whether or not manometric abnormalities are diagnostic of gastroesophageal reflux disease. Quite the contrary, there has been no demonstration that the detection of any manometric finding, or combination of findings, predicts the appropriateness of a particular therapeutic regimen.[18] Therefore medical therapy is empirically selected, based upon symptom characteristics and the extent of mucosal damage seen on an endoscopic examination. With respect to discriminating reflux patients with ENT manifestations versus those without, no study has shown a difference in any manometric parameter including UES pressure, esophageal length, LES pressure, or peristaltic function between these two groups of patients.[4,19] Thus, manometric findings have no proven influence on the diagnosis, staging, or pharmacological treatment algorithms related to gastroesophageal reflux disease with or without otolaryngologic manifestations.

# pH MONITORING FOR ENT MANIFESTATIONS OF GERD

Two major problems emerge when ambulatory pH measurement is used to evaluate patients with suspected otorhinolaryngologic manifestations of GERD: the paucity of relevant normative data and technical limitations. Quantitative analyses of acid reflux have been developed with reference to the distal esophageal mucosa that is continuously bathed by swallowed bicarbonate-containing saliva and, hence, is capable of combating acidification. On the other hand, as far as we know, the larynx is without an efficient defense mechanism against acid injury and experimental data suggest that even a single brief exposure of the true vocal cords to acid daily, or on alternate days, results in the development of a contact ulcer.[20] Similarly, subglottic stenosis can be experimentally induced in as little as seven days by exposing the cartilage beneath a tracheal mucosal injury to gastric acid for a minute a day.[21] Thus, it is not surprising that the quantitative aspects of esophageal acid exposure time determined in the distal esophagus have little bearing on laryngeal symptomatology or pathology. Indeed, studies done placing a pH electrode 5 cm above the LES, have detected no abnormalities in patients with dysphonia[22] or biopsy proven posterior laryngitis.[23]

Another problem with the methodology of pH recording used in the esophagus is that the technique itself is not readily applicable to the hypopharynx or larynx. The physical requirements for intraluminal pH recording dictate that a relatively constant electrical impedance be maintained between the intraluminal probe and the reference skin electrode. The environment of the hypopharynx is such that the electrode would be intermittently dry and moist, vastly altering the impedance of the circuit. As the impedance of the circuit varies, so does the registered pH value, implying that registered changes can be indicative of either true acidification or of electrode drying. This difficulty probably accounts for the wandering baseline often observed in hypopharyngeal recordings as well as the "pseudoreflux" described in investigations that have attempted hypopharyngeal recordings.[8,16] Thus, with existing methodology, recordings free of technical artifact can only be obtained from within the esophagus.

## Intraesophageal Distribution of Reflux

Since it is apparent that pH recordings from the distal esophagus are not discriminating in the detection of ENT manifestations of GERD,[4] the problem arises of where to position the electrode. Different investigators have taken differing approaches to this problem with the general finding that, for whatever probe position selected, there is a greater mean frequency of proximal reflux among the groups with ENT symptomatology. However, one constant among investigations is the concurrent use of multiple pH electrodes within the esophagus. It seems that there is agreement that interpretation of any proximal pH recording is very difficult without a concurrent distal tracing. Two issues are resolved by the distal tracing: whether or not the individual has pathologic reflux and whether or not events recorded proximally occur in association with distal mucosal acidification. One group of investigators has even argued that by applying three criteria to each apparent pH drop in a hypopharyngeal recording (pH drop of >3 pH units, pH drop to a value <5, and pH drop during period of concurrent esophageal acidification), these technically flawed recordings can be used to detect true instances of pharyngeal regurgitation (Figure 12.2).[24] Applying these criteria to recordings from 19 patients with typical symptoms of pharyngeal regurgitation, 26 of 2684 recorded pharyngeal pH drops (1%) were distinguished as true episodes of regurgitation.

The requirement of multiple pH sensing sites for the meaningful assessment of proximal reflux makes the ambulatory pH study technically more complicated. Data loggers and the associated software need multichannel capability. One suitable device is the Synectics Digitrapper (Synectics Medical, 1425 Greenway Drive, Irving, TX 75038). Even more important is the issue of electrode design. Although it is technically feasible to pass several single electrodes as some investigators have done,[19] it is more practical to pass a single electrode assembly with multiple pH recording sites. The study population in question often has an irritated, sensitive throat to begin with, and it is difficult enough for them to tolerate 24 hour placement of a single fine caliber electrode assembly, let alone multiple bulky electrodes. Furthermore, the more comfortable the patient is during

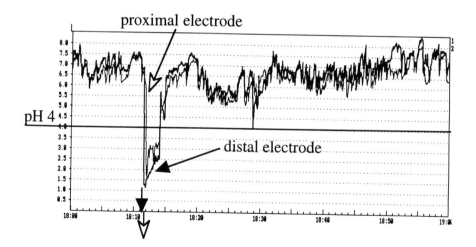

**Figure 12.2.** Example of a dual channel pH probe tracing with the proximal electrode in the hypopharynx and the distal electrode 5 cm above the LES demonstrating a regurgitant event. Open arrows denote proximal acid exposure; filled arrows denote distal acid exposure. Note that the proximal pH drop occurs during a distal pH drop, that the proximal pH drops to less than pH 5, and that the pH fall is > 3 pH units. In a recent investigation these criteria were met for fewer than 1 percent of recorded decreases in pharyngeal pH.[24]

the study the more likely they are to engage in a normal spectrum of daily activities which is crucial to gaining meaningful data. Currently, only two electrode designs have the capability for multisite recording assemblies: antimony monocrystaline sensors or, more recently, an ion sensitive field effect transistor (ISFET) design (Synectics Medical, 1425 Greenway Drive, Irving, TX 75038). Each of these can be configured with multiple pH sensing sites in a small caliber catheter. The ISFET design has the added advantage of exhibiting very rapid response to pH changes (6 pH units within a fraction of a second) and less calibration drift compared to the antimony crystals. Preliminary data have shown that these attributes of the ISFET electrodes permit quantification of another potentially relevant parameter of reflux, the velocity of movement of the refluxate within the esophagus.[25] The relevance of this measurement to otolaryngologic presentations of GERD, however, remains to be determined.

Normal values of intraesophageal acid exposure are dependent upon electrode position within the esophagus. As discussed elsewhere in this book, electrode position is best referenced to the manometrically localized LES, with the convention being to position the electrode 5 cm above the proximal margin of the sphincter. At present, there is no such convention for the evaluation of more proximal reflux, so each laboratory has developed its own set of controls for the electrode placement that they utilize. The importance of consistency in electrode placement and the development of laboratory controls is emphasized by an analysis of the effect of electrode position which concluded that it was these factors, rather than the electrode position per se, that was critical in being able to optimally discriminate control from patient populations.[26] The shortcoming of this situation is that it is hard to compare data among laboratories not using the same electrode position. Some help was provided by a recent study examining the normal distribution of refluxate within the esophagus utilizing a five electrode ISFET assembly with pH recordings 3, 6, 9, 12, and 15 cm proximal to the LES.[25] Figure 12.3 illustrates the key findings from this

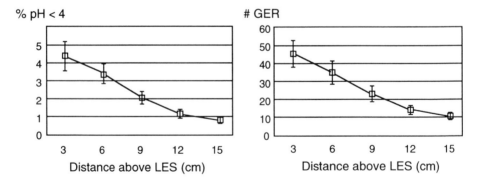

**Figure 12.3.** Intraesophageal distribution of reflux in 8 control subjects without history or symptoms of reflux disease. Concurrent recordings were obtained from 5 intraesophageal locations using an ISFET electrode assembly. The panel on the left illustrates the acid exposure time while the panel on the right illustrates the number of reflux episodes (mean $\pm$ SE). Note that although reflux events were detected at all esophageal locations, both acid exposure time and the number of reflux events is extremely position dependent, decreasing with increasing distance from the LES. Modified from: Weusten BL, Akkermans LM, vanBerge-Henegouwen GP, Smout AJ. Spatiotemporal characteristics of physiological gastroesophageal reflux. *Am J Physiol* 1994;266:G357–362 with permission.

study, showing a linear decrease in both the number of reflux events recorded and the period of esophageal acid exposure with progressively proximal electrode positions. When the results of other investigators normative data are interpreted relative to the positioning of their pH electrode on this nomogram, there is actually good agreement on normal values.[19,27,28] For an electrode position immediately distal to the upper esophageal sphincter, these data suggest that the normal range for acid exposure is 0–1% of a 24 hour period.

## Using Ambulatory pH Studies to Manage Patients with ENT Related GERD

As is readily evident by the above discussion of the array of symptomatology associated with ENT related GERD, these symptoms also have other potential etiologies and it would be of great clinical value to distinguish reflux related cases with pH monitoring. But here lies a subtle distinction. Despite the data detailed below suggesting differences in the proximal esophageal pH profile of *groups* of GERD subjects with ENT related GERD, no attempt has been made to *prospectively* identify affected patients on the basis of pH recordings.[29] To date, the diagnosis of ENT related GERD has been largely based on excluding other potential entities by history, examination, and treatment trials. Thereafter, the pH study is used to either provide additional supportive evidence of a GERD etiology or to verify the adequacy of antireflux therapy.

Several investigators have utilized proximal esophageal pH recordings to characterize groups of GERD patients with and without ENT symptoms.[4,8,16,19,27,30] Table 12.2 summa-

TABLE 12.2. REPRESENTATIVE DATA FROM AMBULATORY pH STUDIES DONE WITH A PROXIMAL pH ELECTRODE DEMONSTRATING THE HETEROGENEITY OF INCLUSION CRITERIA, ELECTRODE POSITIONING, AND FINDINGS AMONG STUDIES

| Patient Population | n | Nl range | Electrode Position | Reference |
|---|---|---|---|---|
| Asymptomatic normals | 26 | 0–0.9% | proximal esophagus | Dobhan 1993[67] |
| GERD with normal distal pH* | 20 | 0–0.7% | proximal esophagus | Dobhan 1993[27] |
| GERD with abnormal distal pH* | 23 | 0–6.7% % abnormal | proximal esophagus | Dobhan 1993[27] |
| GERD without pharyngeal symptoms | 14 | 7% | proximal esophagus | Jacob 1991[4] |
| GERD without pharyngeal symptoms | 19 | 0% | pharyngeal | Shaker 1995[19] |
| "Pharyngolaryngeal symptoms" | 12 | 0% | pharyngeal | Shaker 1995[19] |
| "Pharyngolaryngeal symptoms" | 14 | 21% | proximal esophagus | Jacob 1991[4] |
| Hoarseness | 10 | 70% | pharyngeal | Katz 1990[30] |
| Posterior laryngitis | 15 | 20% | pharyngeal | Weiner 1989[19] |
| Posterior laryngitis | 10 | 50% | proximal esophagus | Jacob 1991[4] |
| Posterior laryngitis | 43 | 17% | pharyngeal | Koufman 1991[8] |
| Posterior laryngitis | 14 | 29% | pharyngeal | Shaker 1995[19] |
| Subglottic stenosis | 18 | 56% | pharyngeal | Koufman 1991[8] |
| Laryngeal cancer | 26 | 58% | pharyngeal | Koufman 1991[8] |

*pharyngeal symptoms not specified

rizes the findings obtained from the most proximal pH electrode in each of these studies. The methodologic details, techniques of data interpretation, and definitions of patient groups vary significantly among investigations making it difficult to compare them. None the less, it is evident that proximal esophageal reflux is unusual in the groups without pharyngeal symptoms or findings and that abnormal recordings were obtained in 17–56% of the patients with pathologies likely related to reflux.

# TREATMENT OF ENT RELATED GERD

If posterior laryngitis is the sequel of acid injury, then it should visibly improve with antireflux treatment. Despite this hypothesis, there is remarkable paucity of controlled data regarding the effectiveness of antireflux therapy in treating posterior laryngitis. One trial examined antireflux therapy consisting of ranitidine 150 mg twice a day and general antireflux measures during 12 weeks of therapy. In this small study, one of six patients with hoarseness and GERD experienced improvement in symptoms after the trial period.[31] Postulating that more profound acid suppression may be required, another study used vigorous acid inhibition in 16 patients with persistent posterior laryngitis despite prior therapy with $H_2$ blockers.[6] This cohort received 6–24 weeks of omeprazole 40 mg q.h.s. with 4 patients ultimately receiving 40 mg b.i.d. for 6 weeks due to refractory symptoms. Esophageal and laryngeal symptom questionnaires as well as laryngoscopy and esophagoscopy were performed at study entry and exit. Laryngoscopic scores as well as laryngeal and esophageal symptoms indices were improved with therapy. Furthermore, an exacerbation of symptoms was noted within 6 weeks of discontinuing omeprazole supporting a causal effect of gastric acid on laryngeal symptoms and the need for maintenance acid suppression (analogous to the treatment of esophagitis).

Further support for the effectiveness of antireflux therapy in the treatment of posterior laryngitis is found in the largest published series to date, reporting 233 consecutive patients presenting with signs and symptoms of chronic laryngitis.[7] Once excluding patients with other likely causes of laryngitis such as chronic hyperfunctional voice abuse or smoking, 182 patients were thought to have reflux as a contributing factor to their chronic laryngeal symptoms. They were treated in a stepwise fashion beginning with standard antireflux precautions without $H_2$ blockers, emphasizing the importance of not eating prior to going to bed and elevating the head of the bed. Patients who did not respond to nocturnal reflux precautions were the treated with famotidine 20 mg at bedtime. Patients were considered refractory if they continued to have persistent laryngeal symptoms and inflammation after a total of 12 weeks of conservative antireflux therapy and an 6 additional weeks of famotidine therapy. These remaining patients were treated with omeprazole 20 mg qhs or a progressively escalated dose first to 40 mg q.h.s. and finally to 40 mg b.i.d., based on interval assessments every 6–8 weeks. Of those on standard antireflux therapy for 6 weeks, 93/182 or 51 percent had resolution or significant improvement. Forty-eight of 89 who were treated with 6 weeks of H2 blockers had a satisfactory clinical response but 92% had recurrent symptoms on follow-up after they had stopped treatment. Of the remaining patients who were treated with omeprazole 20 mg qd, 34 (83 percent) had a good response and of seven patients who required higher doses of omeprazole, all had severe laryngoscopic changes on initial examination eg. ulceration or granuloma. Four of these seven required continuing high doses of omeprazole (up to 80 mg q.d.) and five underwent fundoplication and became asymptomatic. These result are summarized in Figure 12.4. As in the Kamel study, ulceration and granuloma formation were relative uncommon findings with the most common laryngeal abnormalities being erythema and granularity of the posterior larynx. Patients with mild posterior laryngitis (mostly erythema) responded well to less intensive regimens such as antireflux precautions and $H_2$ blockers, but almost all patients who did not respond to those regimens eventually responded to omeprazole. Finally, there was a general correlation between the severity of laryngoscopic findings and need for more intensive antireflux therapy with the patients requiring omeprazole having marked erythema, stasis of secretions, and granularity with ulcerations, granulomas and hyperkeratosis.

Another recent and extensive treatment trial reported on 68 patients with laryngeal symptoms suggestive of acid reflux. All subjects were evaluated by a symptom questionnaire, videolaryngoscopy, and computerized acoustic analysis of voice function. Treatment was with omeprazole 20 mg b.i.d. and either antacids pc and 2 hours before sleep or cisapride 10 mg q.i.d. if they had a history of bloating.[15] The patients were then reevaluated after 12 weeks of therapy. Posttherapy symptom scores were compared to pretherapy scores and all categories (except odynophagia) were improved. Similarly, improvement in videolaryngoscopic scores were noted for every finding except granulomata. Acoustic analysis recordings improved in the subset of patients initially complaining of hoarseness. In this subset there was a statistically significant improvement in the acoustic parameters of frequency (jitter), amplitude (shimmer) and modal fundamental frequency range such that the improvement of voice function and acoustic measures preceded the improvement of laryngeal appearance.

Data on the efficacy of antireflux surgery in the treatment of reflux laryngitis is sparse.

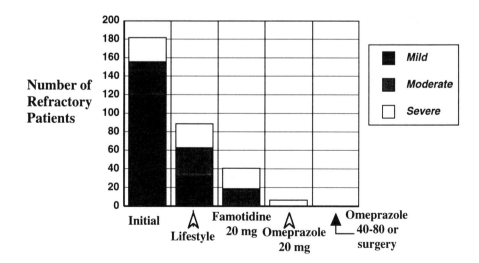

**Figure 12.4.** Resolution of posterior laryngitis with antireflux therapy. Patients were treated in a stepwise fashion for 6 week periods beginning with lifestyle modifications and culminating with high dose omeprazole or antireflux surgery. The number of refractory patients after each intervention is illustrated along with an indication of the severity of their initial laryngoscopic lesions. Note that all patients were eventually resolved and that most of the mild to moderate patients resolved with only lifestyle modifications or low dose famotidine. Drawn from data in Hanson 1995.[7]

One small uncontrolled study reported on 13 patients with chronic persistent laryngeal lesions despite antireflux precautions and an unspecified interval of treatment with $H_2$ receptor antagonists who then underwent Nissen fundoplication. Of the 11 patients available for follow-up, eight demonstrated marked improvement 3 and 6 months postoperatively.[32] Interestingly five of the original 13 had been previously treated for carcinoma of the larynx by either hemilaryngectomy or radiation therapy. A reflux etiology was pursued in these patients because of persistent leukoplakia which improved in three of the five patients after fundoplication.

Evident from the above discussion of available therapeutic data, it is difficult to make definitive statements regarding therapy of posterior laryngitis because the studies are uncontrolled, do not use standardized selection or exclusion criteria, do not uniformly deal with the complicating factors of voice or tobacco abuse, and use a variety of therapeutic regimens. However, despite these limitations and inconsistencies, the results of the three most recent trials suggest the following points: 1) the signs and symptoms of posterior laryngitis do improve with antireflux therapy strengthening the putative relationship between acid reflux and laryngeal irritation; 2) although even simple antireflux measures may be adequate for symptom control, more vigorous acid suppression with a proton pump inhibitor is superior to other therapies for healing laryngitis; 3) pharmacologic treatment of posterior laryngitis often takes longer and requires higher doses of proton pump inhibitors than is typical of esophagitis; 4) severe laryngeal findings are unusual, probably signifying more severe disease and a poorer response to therapy; and 5) the most common laryngoscopic findings due to reflux are subtle—erythema and granularity rather than ulceration or granuloma.

# CASE STUDIES

### Case Study No. 1: "It's a miracle"

A 35 year old woman of northern European origin presented to her family doctor complaining of a 2 year history of intermittent heartburn and belching. Her symptoms occurred several times a week and her symptoms were aggravated by spicy or greasy food, late night meals, and red wine. She denied any difficulties swallowing and denied smoking. She was employed in the chorus of an opera house. Her past medical history was unremarkable and her physical examination was normal. She was prescribed cimetidine 400 mg b.i.d. which she took with relief of her symptoms. After 6 weeks she began taking cimetidine only when needed. Eight months later she noted return of her symptoms with frequent acid regurgitation and nocturnal awakening with coughing after which she would note a change in her voice and occasional hoarseness. She restarted herself on cimetidine, called her family doctor and was referred to a gastroenterologist.

Discussion with the patient revealed that although she had been instructed to avoid late night meals and modify her diet, her occupational lifestyle of evening performances following by large late night meals could not be easily changed. Laryngeal examination revealed swollen, reddish mucosa of the posterior wall of the glottis and an esophagogastroduodenosocopy was normal. She was treated with omeprazole 40 mg q.d. and again instructed to modify her diet and lifestyle. Initially, her symptoms resolved but returned seven weeks later with her reporting two additional incidents of nocturnal awakening and hoarseness necessitating cancellation of several performances. She was very upset explaining that her career as a singer would be over if she could not guarantee her ability to perform.

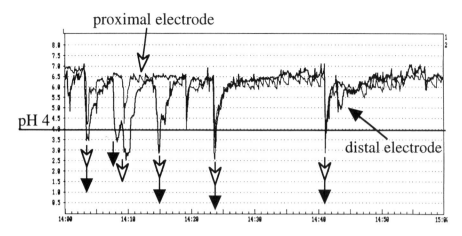

**Figure 12.5.** Dual channel pH probe tracing with the proximal electrode immediately distal to the UES and the distal electrode 15 cm below that point. Open arrows denote proximal acid exposure; filled arrows denote distal acid exposure. This segment of the 24 hour recording demonstrates several episodes of concurrent proximal and distal reflux in this patient with persistent hoarseness.

Twenty-four hour pH monitoring performed off of medications revealed only mild quantitative changes in total distal esophageal acid exposure (7 percent but did show 1.4 percent proximal esophageal acid exposure (Figure 12.5). She was subsequently referred to a surgeon who performed a laparoscopic Nissen fundoplication without complication. The patient was taken off all medications with no further symptoms. She says now, "For me, it's a miracle".

### Case Study No. 2: "Nothing lasts forever."

A 45 year old woman employed in a public relations position of a large international charitable organization complained of chronic throat pain and hoarseness which greatly limited her ability to speak on the phone. She had seen gastroenterologists on and off for 5 years and was treated with H$_2$ receptor antagonists despite the absence of endoscopic lesions. At presentation she had a raspy voice, a normal endoscopy, and posterior erythema on laryngoscopy. Esophageal symptomatology was minimal. Omeprazole 40 mg q.h.s. was instituted with an 80 percent improvement in voice function and clear laryngoscopic improvement after 6 weeks. The patient, however, desired to be free of lifestyle limitations and desired a surgical evaluation.

Esophageal manometry was normal and an ambulatory esophageal pH study done after withholding omeprazole for one week showed 5.3 percent distal and 2.3 percent proximal esophageal acid exposure (Figure 12.6). Advised that there was at least a 50 percent probability of significant improvement with antireflux surgery, the patient was referred to a world renowned laparoscopic surgeon who performed an uneventful laparoscopic Nissen fundoplication. She became asymptomatic within 2 months and discontinued all antireflux precautions and antisecretory medications.

Three years later, the patient was visiting her daughter and offered to walk the dog, a 95 lb male Golden Retriever named Sundance. In the course of their walk, Sundance became interested in a female dog running in the other direction and bolted in pursuit, literally lifting the patient off her feet and, consequently, throwing her, mid section first into the trunk of a

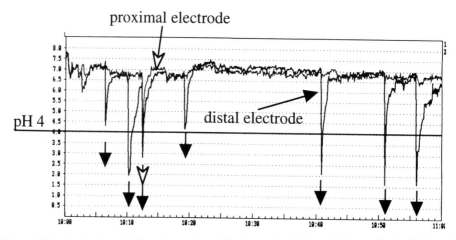

**Figure 12.6.** Dual channel pH probe tracing with the proximal electrode immediately distal to the UES and the distal electrode 15 cm below that point. Open arrows denote proximal acid exposure; filled arrows denote distal acid exposure. This segment of the 24 hour recording demonstrates several episodes of distal reflux, one of which is associated with proximal reflux.

tree. Fortunately, no serious injury was sustained. However, within one week the patient again began having throat pain and hoarseness. An upper GI X-ray demonstrated partial herniation of the stomach through the diaphragmatic hiatus, although the fundoplication itself was intact. The patient was put back on omeprazole 20 mg h.s. and, again, this provided 80% symptom improvement. She is contemplating having her fundoplication redone.

### Case Study No. 3: "Nothing is perfect."

This patient was a 23 year old waitress, struggling to pursue a career as an opera singer. Although clearly talented, her auditions never came off quite well enough because her voice would unpredictably weaken. Laryngoscopic examination by a New York voice specialist suggested mild posterior laryngitis. She had minimal esophageal symptomatology, experiencing heartburn about once a month. None the less, having heard of other performers having had this difficulty cured by antireflux medication, she pursued gastroenterological advice. Endoscopy and an esophageal manometry were normal. An ambulatory pH study done with the patient off of antisecretory medications revealed 5.5 percent distal and 1.3 percent proximal acid exposure. She had no symptomatic response to 40 mg q.h.s. omeprazole. A repeat ambulatory pH study done on this therapy showed 1.2 percent distal and 0.2 percent proximal reflux. The patient was advised that, although it was impossible to be certain, it seemed that there was at best a 50 percent chance of significant improvement with antireflux surgery. Driven by her ambitions, the patient accepted this judgment and went ahead with a laparascopic Nissen fundoplication. The operation went smoothly, but she had no significant improvement in voice function afterward. An ambulatory pH study done 2 months postoperatively showed 0.8 percent distal and no proximal acid exposure with the patient off of all antisecretory medications. She continues to work as a waitress.

These cases typify the emerging observations that the pattern of reflux with otolaryngologic manifestations is different than that of the typical reflux patient. Quantitative analysis of acid reflux may detect minimal abnormalities, yet antireflux surgery may effectively cure a patient. On the other hand, surgical repair cannot be regarded as definitive treatment for every patient. Surgical interventions are subject to "wear and tear" with their own inherent susceptibility to complications or breakdown. Sometimes, however, when an individual's profession is dependent upon a flawless voice, choosing a "dramatic solution" may be the only choice. Furthermore, predicting outcome in this very diverse group of patients is difficult, since laryngeal symptoms and mild laryngoscopic abnormalities are potentially multifactorial in etiology. Finally, judgments of improvement in vocal characteristics or subtle laryngeal pathology is highly subjective.

# SUMMARY

Partly as a result of vast improvements in the pharmacological treatment of esophagitis, the past decade has seen substantial interest focused on the extraesophageal manifestations of GERD. Case reports and uncontrolled series suggest a broad spectrum of pharyngeal and laryngeal symptomatology and pathology attributable to reflux. The best characterized cases are of posterior laryngitis and tracheal stenosis. Much, however, remains to be learned. Even these better characterized entities are often multifactorial in pathogenesis with reflux being just one relevant factor. Esophageal pH monitoring has emerged as the best tool for further elucidating this relationship. However, it is already apparent that the standard methodology for performing the pH examination has little relevance to the ENT spectrum of the disease and methodologic modifications are necessary. In particular, multichannel studies utilizing proximal esophageal or pharyngeal pH electrodes are required. Moreover, much more investigative work is necessary before the results of these studies can be reliably interpreted. Until then, pH monitoring is a useful evaluative tool but does not obviate the need for a careful history and examination and even in the best of circumstances, can still be misleading. Similarly, significant questions remain to be addressed regarding the prevalence of acid injury above the esophageal inlet, the specificity of the pharyngeal and laryngoscopic findings for acid injury, and the responsiveness of these lesions to antireflux medication.

# REFERENCES

1. von Leden H, Moore P: Contact ulcer of the larynx, experimental observations. *Arch Otolaryngol* 72:746–751, 1960.
2. Jackson C: Contact ulcer of larynx. *Ann Otol Rhinol Laryngol* 37:227–230, 1928.
3. Cherry J, Margulies SI: Contact ulcer of the larynx. *Laryngoscope* 78:1937–1940, 1968.
4. Jacob P, Kahrilas PJ, Herzon G: Proximal esophageal pH-metry in patients with 'reflux laryngitis'. *Gastroenterology* 100:305–310, 1991.
5. Kambic V, Radsel Z: Acid posterior laryngitis, aetiology, histology, diagnosis and treatment. *J Laryngol Otol* 98:1237–1240, 1984.
6. Kamel PL, Hanson D, Kahrilas PJ: Prospective trial of omeprazole in the treatment of posterior laryngitis. *Amer J Med* 96:321–326, 1994.
7. Hanson DG, Kamel PL, Kahrilas PJ: Outcomes of anti-reflux therapy in the treatment of chronic laryngitis. *Ann Otol Rhinol Laryngol* 104:550–555, 1995.
8. Koufman JA: The otolaryngologic manifestations of gastroesophageal reflux disease. *Laryngoscope* 10(supp 53):1–78, 1991.

9. Cote DN, Miller RH: The association of gastroesophageal reflux and otolaryngologic disorders. *Comp Ther* 21:80–84, 1995.

10. Gaynor EB: Gastroesophageal reflux as an etiologic factor in laryngeal complications of intubation. *Laryngoscope* 98:972–979, 1988.

11. Wydner EL, Stellman SD: Comparative epidemiology of tobacco-related cancers. *Cancer Res* 37:4608–4622, 1977.

12. Olsen NR: Effects of stomach acid on the larynx. *Proc Am Laryngol Assoc* 104:108–112, 1983.

13. Morrison MD: Is chronic gastroesophageal reflux a causative factor in glottic carcinoma? *Otolaryngol Head Neck Surg* 99:370–373, 1988.

14. Ward PH, Hanson DG: Reflux as an etiologic factor of carcinoma of the laryngopharynx. *Laryngoscope* 98:1195–1199, 1988.

15. Shaw GY, Searl JP: Laryngeal manifestations of gastroesophageal reflux before and after treatment with omeprazole. *South Med J* 1996, In Press.

16. Weiner GJ, Koufman JA, Wu WC, Cooper JB, Richter JE, Castell DO: Chronic hoarseness secondary to gastroesophageal reflux disease: documentation with 24-H ambulatory pH monitoring. *Am J Gastroenterol* 84:1503–1508, 1989.

17. Kahrilas PJ: GERD and its complications. In: Sleisenger MH, Fordtran JS; Editors. *Gastrointestinal Disease: Pathophysiology, Diagnosis, Management, 6th ed.* Philadelphia: WB Saunders, In Press.

18. Kahrilas PJ, Clouse RE, Hogan WJ: American Gastroenterological Association technical review on the clinical use of esophageal manometry. *Gastroenterology* 107:1865–1884, 1994.

19. Shaker R, Milbrath M, Ren J, et al: Esophagopharyngeal distribution of refluxed gastric acid in patients with reflux laryngitis. *Gastroenterology* 109:1575–1582, 1995.

20. Delahunty JE, Cherry J: Experimentally produced vocal cord granulomas. *Laryngoscope* 78:1941–1947, 1968.

21. Little FB, Koufman JA, Kohut RI, Marshall RB: Effect of gastric acid on the pathogenesis of subglottic stenosis. *Ann Otol Rhinol Laryngol* 94:516–519, 1985.

22. Kjellén G, Brudin L: Gastroesophageal reflux disease and laryngeal symptoms. Is there a causal relationship? *ORL* 56:287–290, 1994.

23. Wilson JA, White A, vonHaacke NP, et al: Gastroesophageal reflux and posterior laryngitis. *Ann Otol Rhinol Laryngol* 98:405–410, 1989.

24. Williams RB, Ali GN, Wallace KL, deCarle DJ, Cook IJ: Detection of acid regurgitation with a pharyngeal pH electrode: validation of measurement criteria. *Gastroenterology* 108:A258 (abstract), 1996.

25. Weusten BL, Akkermans LM, vanBerge-Henegouwen GP, Smout AJ: Spatiotemporal characteristics of physiological gastroesophageal reflux. *Am J Physiol* 266:G357–362, 1994.

26. Singh P, Taylor RH, Colin-Jones DG: Simultaneous two level oesophageal pH monitoring in healthy controls and patients with oesophagitis: comparison between two positions. *Gut* 35:304–308, 1994.

27. Dobhan R, Castell DO: Normal and abnormal proximal esophageal acid exposure: results of ambulatory dual-probe pH monitoring. *Am J Gastroenterol*, 88:25–29, 1993.

28. Harding SM, Richter JE, Guzzo MR, Schan CA, Alexander RW, Bradley LA: Asthma and gastroesophageal reflux: acid suppressive therapy improves asthma outcome. *Am J Med* 100:395–405, 1996.

29. Kahrilas PJ, Quigley EMM: Clinical esophageal pH monitoring, a technical review for practice guideline development. *Gastroenterology* 110:1982–1996, 1996.

30. Katz PO: Ambulatory esophageal and hypopharyngeal pH monitoring in patients with hoarseness. *Am J Gastroenterol* 85:38–40, 1990.

31. McNally PR, Maydonovitch CL, Prosek RA, Collette RP, Wong RK: Evaluation of gastroesophageal reflux as a cause of idiopathic hoarseness. *Dig Dis Sci*, 34:1900–1904, 1989.

32. Deveney CW, Benner K, Cohen J: Gastroesophageal reflux and laryngeal disease. *Arch Surg* 128:1021–1027, 1993.

# 13

# Pulmonary Abnormalities in Gastroesophageal Reflux Disease

**Susan M. Harding, M.D.**

Gastroesophageal reflux disease (GERD) can have a major impact on pulmonary disease. The association between GERD and pulmonary disease is best characterized in asthma. The first clinical observation of the association between asthma and GERD was made by Sir William Osler who stated that asthmatics "learn to take their large daily meal at noon in order to avoid nighttime asthma which occurred if they ate a full supper."[1] Since Osler's time, data suggests that therapy of GERD in asthmatics not only improves asthma symptoms and pulmonary function, but also decreases prednisone medication usage.[2,3,4,5] Ambulatory 24-hour esophageal pH monitoring is a useful diagnostic tool especially in asthmatics who deny esophageal reflux symptoms.[3] This chapter will discuss the prevalence, pathogenesis, diagnosis, and treatment of GERD-associated asthma.

## PREVALENCE OF GERD IN ASTHMATICS

The prevalence of GERD among asthmatics is estimated to be between 34 percent and 89 percent.[6] Esophageal reflux symptoms are common in asthmatics. Using a GERD symptom questionnaire in 109 asthmatics, Field et al. noted that 77 percent of asthmatics reported heartburn, 55 percent complained of regurgitation, and 24 percent experienced swallowing difficulties.[7] In the week prior to completing the questionnaire, 41 percent of asthmatics noted reflux-associated respiratory symptoms and 28 percent used their inhalers while experiencing GERD symptoms.[7] Importantly, some asthmatics have significant GERD without esophageal symptoms. In difficult-to-control asthmatics GERD was clinically "silent" in 24 percent.[3]

Asthmatics also have a high incidence of esophageal motility abnormalities, esophagitis, and abnormal esophageal acid contact times. In 97 consecutive asthmatics, Kjellen et al. found that 38 percent had evidence of esophageal dysmotility and 27 percent had lower

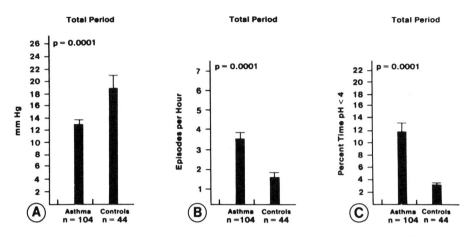

**Figure 13.1.** Lower esophageal sphincter pressure (Panel A) and reflux parameters (Panels B and C) in asthmatics and control subjects. Compared to controls, asthmatics had significantly lower lower esophageal sphincter pressures, more frequent reflux episodes, and higher esophageal acid contact times. (From Sontag SJ et al., *Gastroenterology* 1990, 99:666).

esophageal sphincter (LES) hypotension.[8] Sontag evaluated 186 consecutive adult asthmatics with endoscopy and esophageal biopsy finding that 43 percent had evidence of esophagitis or Barrett's esophagus.[9] Using 24-hour esophageal pH monitoring and esophageal manometry in 104 consecutive asthmatics and 44 controls, the same group observed that 82 percent of asthmatics had higher than normal esophageal acid contact times.[10] Also, the asthmatics, compared to controls, had significantly lower LES pressures and more frequent reflux episodes (Figure 13.1).[10] Multiple studies show that GERD is quite prevalent in the asthma population and GERD needs to be considered as an asthma trigger in every asthmatic.

# MECHANISMS OF GERD RELATED BRONCHOCONSTRICTION

There are three potential mechanisms whereby esophageal acid triggers airflow obstruction in asthmatics. These include a vagally mediated reflex, where acid in the esophagus stimulates acid sensitive receptors resulting in bronchoconstriction; heightened bronchial reactivity; and microaspiration of gastric contents into the upper airway.

## Evidence for a Vagal Reflex

Many studies support a vagal mechanism. In a dog model, esophageal acid caused an increase in respiratory resistance which was ablated with bilateral vagotomy.[11] In human studies, Mansfield et al. observed in asthmatics with reflux that total respiratory resistance increased 10 percent with the infusion of acid into the esophagus.[12] In another study, Wright et al. monitored 136 subjects before and after esophageal acid infusions and noted reductions in airflow and arterial oxygen saturation with esophageal acid, which were not present with atropine pretreatment.[13] However, others report conflicting results, including a study of 15 nocturnal asthmatics, where esophageal acid caused no significant change in airflow resistance.[14]

We performed a series of studies evaluating the pathogenesis of esophageal acid induced bronchoconstriction. Using esophageal infusions of normal saline and acid, we showed that peak expiratory flow rates (PEF) decreased with esophageal acid in normal controls, asthmatics with GERD, asthmatics without GERD, and subjects with GERD alone.[15] Esophageal acid clearance resulted in improvement in PEF in all groups except in the asthma with GERD group (Figure 13.2). This bronchoconstrictor effect was not dependant on proximal esophageal acid exposure, a prerequisite for microaspiration.[15] Subsequently, we infused esophageal acid while the subjects remained in the supine position. Again, esophageal acid caused a decrease in PEF and an increase in specific airway resistance in asthmatics with GERD, which didn't improve despite esophageal acid clearance.[16] Vagolytic doses of atropine given before esophageal infusions partially ablated the bronchoconstrictor response, supporting the role of a vagally mediated reflex.[17]

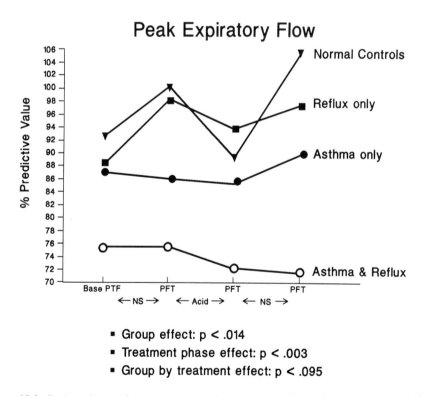

**Figure 13.2.** Peak expiratory flow rates expressed as percent predicted during esophageal infusion of normal saline and acid in normal controls, patients with reflux only, asthma only, and asthma and reflux. All groups had a decrease in peak expiratory flow rate with esophageal acid. Note that esophageal acid clearance resulted in improvement in peak expiratory flow rate in all groups except those with asthma and reflux. Base-baseline; PFT—pulmonary function test; NS—normal saline. (From Schan CA et al., *Chest* 1994; 106:734).

### Heightened Bronchial Reactivity

This mechanism hypothesizes that GERD aggravates asthma by increasing the bronchomotor response to other stimuli. This was shown experimentally by Herve et al. who examined measures of airway reactivity (voluntary isocapnic hyperventilation of dry air and methacholine challenge tests) after esophageal saline and acid infusions.[18] Esophageal acid caused a two-fold increase in bronchoconstriction after voluntary isocapnic hyperventilation. Also, the total dose of methacholine required to reduce the $FEV_1$ by 20 percent ($PD_{20}$) was significantly lower with acid infusion versus saline. Furthermore, this bronchoconstrictor response to esophageal acid was abolished with atropine pretreatment, showing that the vagus nerve also plays a role in the heightened bronchial reactivity model.[18]

### Microaspiration

Microaspiration of esophageal acid into the upper airway can cause significant bronchospasm. Convincing experiments were performed by Tuchman et al. in a cat model. They infused 10 ml of acid into the esophagus noting a 1.5-fold increase in total lung resistance compared to a five-fold increase with 0.5 ml of acid infused into the trachea.[19] Furthermore, the esophageal acid bronchoconstrictor response occurred in only 60 percent of the cats versus 100 percent of the cats with tracheal acid.[19] Interestingly, the affect of tracheal acid on total lung resistance was abolished after bilateral surgical vagotomy, so even in the aspiration model, the vagus nerve plays a role.[19]

The importance of microaspiration was also shown in humans. Jack et al. monitored simultaneous tracheal and esophageal pH in four severe asthmatics.[20] Esophageal acid caused a 8 liter a minute decrease in peak expiratory flow rate. However, if both esophageal and tracheal acid were present, peak expiratory flow rates decreased an average of 84 liters a minute.[20] So, episodes of tracheal microaspiration were associated with a significant deterioration in pulmonary function.[20]

In summary, esophageal acid causes bronchoconstriction by a vagally mediated reflex, however, if microaspiration is present there is further augmentation of this bronchoconstrictor response (Figure 13.3). It is interesting that the vagus nerve plays a role in all three mechanisms.

# MAKING THE DIAGNOSIS
# OF GERD IN ASTHMATICS

## Role of 24 Hour pH Testing

All asthmatics should be questioned about GERD symptoms. Pertinent historical questions are illustrated in Table 13.1 and include esophageal and extraesophageal manifestations of GERD. If the patient's history is consistent with GERD, no further diagnostic studies are necessary and a trial of aggressive anti-reflux therapy should be started. Further testing is recommended in patients in whom emperic therapy of GERD is unsuccess-

# Esophageal Acid Induced Bronchoconstriction

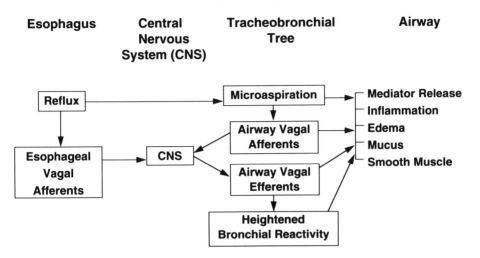

**Figure 13.3.** Mechanisms of esophageal acid induced bronchoconstriction: vagally mediated reflex, heightened bronchial reactivity and microaspiration. The vagus nerve plays a role in all three mechanisms resulting in bronchoconstriction and airway inflammation.

## TABLE 13.1. SYMPTOMS OF GERD IN ASTHMATICS

Esophageal Symptoms
  Heartburn
  Regurgitation
  Waterbrash
  Dysphagia
Extraesophageal Symptoms
  Sore Throat
  Choking
  Hoarseness
  Dental Erosion
  Chest Pain
  Cervical Pain
Worsened Asthma Symptoms With
  Eating
  Alcohol
  Reflux Symptoms
  Supine Position
  Theophylline
  Systemic B-adrenergic Agents
Using an Inhaler While Experiencing GERD Symptoms
Clinically Silent

ful or in those patients who have symptoms suggesting complicated GERD (esophagitis, esophageal stricture, Barrett's esophagus, or neoplasm). In patients with symptoms consistent with complicated GERD, endoscopy should be considered because it provides direct visualization of the esophageal mucosa and biopsies can be obtained.[21]

Careful questioning about GERD is not enough to identify all patients with asthma triggered by GERD. Asthma symptoms may be the sole manifestation of GERD, with many patients denying heartburn and regurgitation. Irwin et al. studied a group of difficult-to-control asthmatics finding that GERD was clinically "silent" in 24 percent.[3] Twenty four-hour esophageal pH testing identified these patients. Since antireflux therapy improved their patient's asthma, Irwin et al. recommended 24-hour esophageal pH testing in all asthmatics who are difficult-to-control or who are on chronic prednisone therapy.[3] More recently, the American Gastroenterological Association's Medical Position Statement on the Clinical Use of Esophageal pH Recording recommended testing asthmatics suspected of having reflux triggered asthma.[22,23]

The initial esophageal pH study, especially when employing the dual pH probe, may be prognostic as well as diagnostic in patients with asthma and suspected GERD. As discussed in Chapter 12, no consensus exists for the location for placing the proximal probe, but this is the best test for diagnosing possible microaspiration which might be important

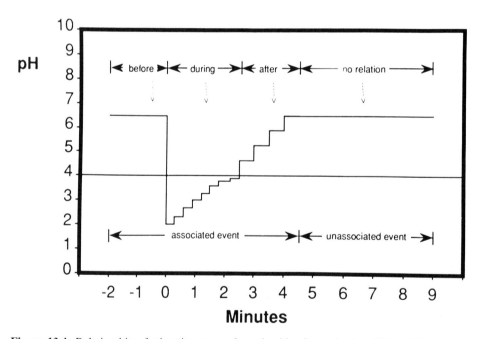

**Figure 13.4.** Relationship of wheezing to esophageal acid reflux episodes. Of the 142 wheezing events, 21 percent occurred before, 16 percent occurred during and 10 percent occurred after the reflux episode. Fifty four percent of the wheezing events had no relation to the reflux episodes. (From Sontag SJ, et al.: *Ambulatory Esophageal pH Monitoring: Practical Approach in Clinical Applications*, JE Richter and DO Castell, eds, Igaku-Shoin, New York, 1991, p161).

in these patients' symptoms. For example, one retrospective study[24] found that 24/34 (71%) of patients with abnormal reflux by 24-hour pH testing, including five with proximal acid reflux only, achieved good to excellent response of their pulmonary symptoms after aggressive antireflux treatment. In contrast, the five patients who received antireflux treatment despite having no documentable reflux, showed no response of their pulmonary symptoms. In another study, a history of frequent regurgitation or excessive proximal esophageal reflux just below the upper esophageal sphincter were the best predictors of acid response to aggressive omeprazole therapy[2].

Twenty four-hour esophageal pH testing also allows correlation of asthma symptoms with reflux episodes. Sontag et al. analyzed esophageal pH tracings of 48 consecutive asthmatics.[25] A wheezing event was considered to be associated with GERD if wheezing occurred during the reflux episode or within the two minute period before or after the reflux episode (Figure 13.4). Of the 142 wheezing events, 20 percent occurred before, 16 percent occurred during, and 10 percent occurred after the reflux episode. Fifty four percent of wheezing events had no time relationship to reflux episodes.[25] Finally, esophageal pH testing allows documentation of adequate acid suppression with therapeutic interventions. Therefore, 24-hour esophageal pH testing plays a key role in diagnosing GERD in asthmatics.

## Asthma Outcome With Antireflux Medical Therapy Trials

Since esophageal acid is a trigger of asthma, aggressive antireflux therapy should improve asthma symptoms and pulmonary function. In difficult to control asthmatics requiring more than 10 mg of prednisone Q.O.D., antireflux therapy and inhaled corticosteroid use were the two most helpful interventions which converted difficult-to-control asthmatics into ones that were easier to control.[3] However, previous medical trials using antacids or $H_2$-receptor antagonists at standard doses have reported only mixed results as illustrated in Table 13.2[26-31] Recently, two placebo-controlled crossover studies using the proton pump inhibitor omeprazole examined asthma outcome in asthmatics with GERD. Meier et al. using omeprazole 20 mg twice a day for six weeks noted that 29 percent of asthmatics with GERD had omeprazole responsive asthma.[32] When evaluating the nonresponders, 45 percent still had esophagitis on omeprazole. The nonresponders also had three to five times higher esophageal acid contact times than the responders. Ford et al. examined asthmatics with GERD comparing four weeks of omeprazole, 20 mg a day, to placebo in a crossover study.[33] They found no significant difference in asthma symptoms or peak expiratory flow rates.[32] Shortcomings of these two studies include inadequate control of esophageal acid and the study duration was too short to adequately address asthma outcome.

We performed a trial where daily asthma symptoms, peak expiratory flow rates and spirometry were assessed before and after three months of acid suppressive therapy with omeprazole.[2] Asthmatics with GERD and a greater than 20 percent improvement in asthma symptoms and/or a 20 percent improvement in peak expiratory flow rates with three months of acid suppressive therapy were considered asthma responders by a priori definitions. Seventy three percent were asthma responders. Responders reduced their asthma symptoms by 57% and improved their peak expiratory flow rates by 9% and had significant improvement in pulmonary function tests including the $FEV_1$, forced expira-

TABLE 13.2. MEDICAL TRIALS OF GERD-RELATED ASTHMA*

| Author | Number Patients | Number Controls | Tx | Asthma Outcome |
|---|---|---|---|---|
| Kjellen et al.[25] | 31 | 31 | Antacids/Alginic acid | 54% improved |
| Goodall et al.[26] | 18 | Placebo crossover | Cimetidine 200 mg QID for 6 weeks† | Increased PEF[ii] and decreased asthma symptoms in 78% |
| Harper et al.[27] | 15 | — | Ranitidine 150 mg BID for 8 weeks§ | Decreased symptoms and improved PFT's over entire group† |
| Nagel et al.[28] | 15 | Placebo crossover | Ranitidine 450 mg a day for 1 week | No difference |
| Ekstrom et al.[29] | 24 | Placebo crossover | Ranitidine 150 mg BID for 4 weeks§ | Mild decrease nocturnal symptoms and decreased MDI use¶ |
| Depla et al.[30] | 1 | — | Omeprazole 20 mg a day for 3 months | Complete relief of symptoms |
| Meier et al.[31] | 15 | Placebo crossover | Omeprazole 20 mg BID for 6 weeks | 29% increased FEV$_i$** by 20% |
| Ford et al.[32] | 11 | Placebo crossover | Omeprazole 20 mg a day for 4 weeks | No difference |
| Harding et al.[2] | 30 | — | Omeprazole 20–60 mg/day; documented acid suppression | 73% increased PEF[ii] or decreased symptoms by 20% |

*GERD = Gastroesophageal reflux
†QID = Four times
§BID = Twice a day
‡PFT's = Pulmonary function tests
iiPEF = Peak expiratory flow rate
¶MDI = Metered dose inhaler
**FEV$_1$ = Forced expiratory volume at one second

tory flow during the middle half of forced vital capacity ($FEF_{(25-75\%)}$), and peak expiratory flow rate with acid suppressive therapy[2]. Also, nearly one-third of asthmatics with GERD required more than 20 mg of omeprazole a day to control GERD.[2] The time course of asthma symptom improvement in the responders is shown in Figure 13.5. Two variables predicted asthma improvement. The presence of regurgitation at least once a week and/or excessive proximal acid reflux by 24-hour esophageal pH testing predicted asthma improvement with a 100% sensitivity and a positive predictive value of 79%.[2] Thus, documented acid suppression with proton pump inhibitors can improve asthma in nearly 75 percent of asthmatics with GERD. This response rate approximates the asthma improvement rate with anti-reflux surgery.

# Asthma Symptom Score

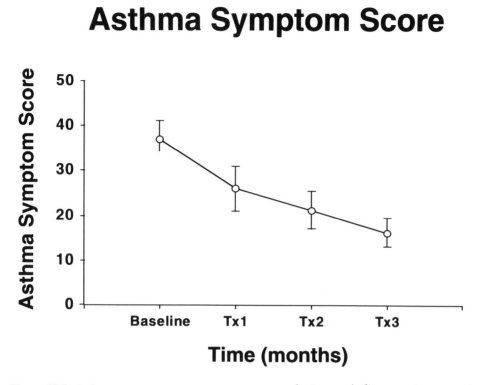

**Figure 13.5.** Asthma symptom score response to omeprazole therapy. Asthma symptom score at baseline and at omeprazole treatment month one (T × 1), treatment month two (T × 2), and treatment month three (T × 3) in asthma responders. Means ± standard error are shown. The asthma responders had a 57 percent reduction in asthma symptoms after three months of acid suppressive therapy. (Data from Harding SM et al.; *Am J Med* 1996; 100:401).

## SURGICAL AND COMBINED MEDICAL AND SURGICAL TRIALS

The largest surgical trial examining asthmatics with GERD was by Perrin-Fayolle et al. who reported five year follow-up in forty four asthmatics.[34] Twenty five percent of asthmatics had total resolution of their asthma symptoms, 41 percent had improvement, and 34 percent noted no change.[33] Combined results of all surgical therapy trials reported to date show that 34 percent of asthmatics were free of asthma symptoms postoperatively, 42 percent were improved and 24 percent were unchanged.[5,34,36,37] Study design flaws include the lack of a control group, poor documentation of airflow obstruction pre and postoperatively and no proof that reflux was controlled in the postoperative state.

Two placebo controlled trials compared medical versus surgical therapy in asthmatics with GERD. Sontag et al. reported a five year follow up comparing ranitidine 150 mg three times a day to surgery (Nissen fundoplication).[5] In the surgically treated group, 75 percent had resolution or improvement in asthma outcome versus 9 percent in the ranitidine group and 4 percent in the control group. Prednisone was discontinued in 33 percent of the surgical group, 11 percent of the ranitidine group, and in 0 percent of the control group.[5]

Larrain et al. compared cimetidine 300 mg Q.I.D. to surgical therapy (Hill procedure) in nonallergic asthmatics with GERD.[4] After six months, asthma medication scores decreased significantly in both treatment groups versus the placebo group.[4] Asthma was considered improved in 76 percent of the surgically treated group, 74 percent of the medically treated group and 36 percent of the placebo group.[4] Unfortunately, neither study used proton pump inhibitors or documented control of acid reflux by 24-hour esophageal pH testing.

# THERAPY OF GERD ASSOCIATED ASTHMA

Since aggressive antireflux therapy can significantly improve asthma outcome in approximately 75 % of asthmatics with GERD, all patients should have a therapeutic trial. All patients should be educated on lifestyle therapy.[32] If possible, medications that decrease LES pressure should be avoided including theophylline.[39] Medical therapy includes antacids, $H_2$ antagonists (cimetidine, ranitidine, nizatidine, and famotidine), proton pump inhibitors (omeprazole and lansoprazole), and prokinetic agents (bethanechol, metoclopramide, and cisapride).[40] Bethanechol is contraindicated in asthmatics because it can induce bronchospasm.[40] Also, metoclopramide has a 20 percent to 50 percent incidence of side effects including fatigue, restlessness, tremor, Parkinsonism, or tardive dyskinesia.[40] Surgical interventions include the Nissen, Toupet, and Belsey fundoplications, and the Hill gastropexy. Newer laproscopic techniques decrease hospital time; however, their long term efficacy is unknown.

Aggressive initial therapy of GERD in asthmatics should include a proton pump inhibitor (omeprazole or lansoprazole) since they provide better acid control than $H_2$ antagonists or prokinetic agents.[41] A trial of omeprazole 20 mg B.I.D. or lansoprazole 30 mg B.I.D. should be initiated and continued for three months. This dose may be necessary because Harding et al. noted that omeprazole 20 mg a day did not suppress esophageal acid in 27 percent of asthmatics with GERD.[6] During the therapeutic trial, patients should monitor their asthma symptoms and daily peak expiratory flow rates. Anti-reflux therapy is considered successful if there is a 20 percent decrease in asthma symptoms or daily corticosteroid dose or a 20 percent improvement in peak expiratory flow rates. If the patient improves, then chronic maintenance therapy should be considered (see next paragraph). If treatment is unsuccessful, then either the patient's asthma is not triggered by GERD and aggressive anti-reflux therapy can be discontinued or GERD was not adequately controlled. In this case, 24-hour esophageal pH testing can be performed while the patient is on antireflux medication. If GERD is not controlled on omeprazole 20 mg B.I.D. then the dose should be increased to 40 mg B.I.D. and a prokinetic agent such as cisapride can be added.

Maintenance therapy can include a proton pump inhibitor or a high dose of $H_2$ antagonist. Prokinetic agents are usually used in combination with acid suppressive drugs. All

**TABLE 13.3. CLINICAL APPROACH TO GERD IN ASTHMATICS**

1) Inquire about GERD symptoms in all asthmatics.
   a) If GERD symptoms are present; begin a therapeutic anti-reflux trial.
   b) If GERD symptoms are absent; consider 24-hour esophageal pH test especially in patients on oral prednisone and/or who are difficult-to-control.
2) If GERD is present, begin a three month trial of omeprazole 20 mg BID or lansoprazole 30 mg BID while monitoring asthma symptoms and peak expiratory flow rates.
3) If asthma is improved, begin maintenance GERD therapy with a high dose $H_2$ antagonist or proton pump inhibitor $+/-$ a prokinetic agent.
4) If asthma is not improved after a three month trial:
   a) Asthma is not triggered by GERD and aggressive anti-reflux therapy can be discontinued, or
   b) GERD is not adequately controlled on the anti-reflux regimen. A 24-hour dual esophageal pH test can be done to document adequate acid suppression.
5) Anti-reflux surgery should be considered in patients who have normal esophageal motility and a hypotonic LES, who are relatively young in age, and don't want to take chronic medications or deal with the possible long term side effects of proton pump inhibitors.

patients requiring a proton pump inhibitor for chronic therapy should have the option of surgery discussed, especially in younger patients, since there are still unanswered questions about the long term safety of proton pump inhibitors.[41,42] Optimal surgical candidates include those with LES hypotension and normal esophageal motility.[43,44] Laproscopic fundoplication is best employed in patients with uncomplicated reflux disease.[45] There are no studies evaluating the cost benefit analysis of medical versus surgical therapy for long term therapy of GERD-associated asthma. Table 13.3 reviews the clinical approach of GERD in asthmatics.

## Case Study

JC is a 57-year-old woman who presented in March of 1994 with steroid-dependent asthma. She denied asthma as a child and an allergy work-up was negative. She complained of dyspnea, wheezing and a nonproductive cough. She reported multiple nocturnal awakenings with active bronchospasm. She complained of daily indigestion and frequent episodes of regurgitation. She denied dysphagia. She noticed that she often used her inhaler after reflux episodes. Her GERD symptoms began approximately one year prior to the onset of her pulmonary symptoms. Medications on initial presentation are listed in Table 13.4 and included prednisone 20 mg a day. Her initial pulmonary function tests showed significant airflow obstruction (Table 13.4).

Esophageal pH monitoring showed significant distal and proximal reflux, especially in the distal probe while in the supine position (Figure 13.6 and Table 13.4). Esophageal endoscopy revealed grade II esophagitis. She was started on omeprazole 20 mg B.I.D., but continued to have reflux symptoms, so a prokinetic medication (cisapride) was added. In May of 1994, she had marked improvement in her asthma symptoms after three months of aggressive antireflux therapy. Her prednisone dose was tapered from 20 mg a day to 5 mg a day (Table 13.4). Repeat 24-hour esophageal pH testing showed adequate acid suppression (Table 13.4). After six months of aggressive antireflux therapy, she was off prednisone and had normal pulmonary function tests.

**TABLE 13.4. CLINICAL COURSE OF JC**

| | March 1994 Initial Presentation | May 1994 On Reflux Medications | January 1995 S/P Toupet Fundoplication |
|---|---|---|---|
| Pulmonary Function Tests | | | |
| FVC (% predicted)* | 1.7 (48%) | 3.00 (83%) | 2.99 (84%) |
| FEV$_1$ (% predicted)** | 1.18 (42%) | 2.38 (88%) | 2.57 (97%) |
| FEV$_1$%† | 69% | 79% | 86% |
| PEF[ii] | 1.29 (23%) | 4.59 (73%) | 5.09 (81%) |
| Manometry | | | |
| LESP[#] | 12 mm Hg | — | 10mm Hg |
| Esophageal pH | | | |
| Distal | | | |
| Total (nl[S] < 5.46%) | 16.4% | 2.0% | 0.7% |
| Upright (nl < 8.05%) | 13.1% | 2.6% | 1.1% |
| Supine (nl < 2.48%) | 21.6% | 0% | 0% |
| Proximal | | | |
| Total (nl < 1.10%) | 1.4% | 0.7% | 0.4% |
| Upright (nl < 1.70%) | 2.2% | 0.9% | 0.5% |
| Supine (nl < 0.6%) | 0.2% | 0% | 0% |
| Asthma Medications | Prednisone 20 mg Q.D. Albuterol M.D.I. Ipratropium M.D.I. Corticosteroid M.D.I. Theophylline | Prednisone 5 mg Q.D. Albuterol M.D.I. Ipratropium M.D.I. Corticosteroid M.D.I. | Corticosteroid M.D.I. Albuterol M.D.I. prn |
| Reflux Medications | — | Omeprazole 20 mg B.I.D. Cisapride 10 mg Q.I.D. | — |

*FVC = Forced vital capacity (liters)
**FEV$_1$ = Forced expiratory volume at one second (liters)
†FEV$_1$% = FEV$_1$/FVC
[ii]PEF = Peak expiratory flow rate (liters/second)
[#]LESP = Lower esophageal sphincter pressure (mm Hg)
[S]nl = Normal
[8]MDI = Metered dose inhaler

JC was concerned about the high cost of chronic medical antireflux therapy and the un-known risks of long-term proton pump inhibitor use. She switched to high dose H$_2$ antago-nist therapy (ranitidine 300 mg B.I.D.) and cisapride, however, her reflux symptoms re-curred. Since she had normal esophageal peristalsis and her asthma clearly improved on acid suppressive antireflux therapy, a surgical evaluation was recommended.

JC underwent a Toupet fundoplication in September of 1994. Four months post-opera-tively (January 1995) she had normal esophageal acid contact times off anti-reflux medica-tion and normal pulmonary function tests (Figure 13.7 and Table 13.4). She denied reflux

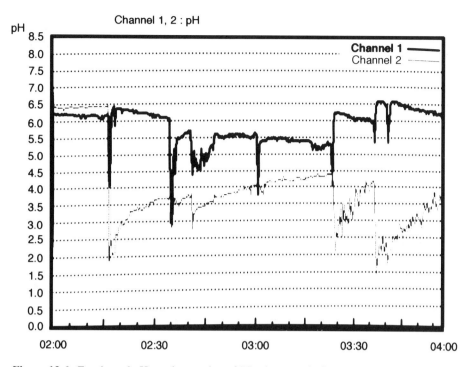

**Figure 13.6.** Esophageal pH monitor tracing of JC prior to antireflux therapy. Dual esophageal pH probe was placed in the distal esophagus (Channel 2, 5 cm above the LES) and proximal esophagus (Channel 1, 2 cm below the UES). Prolonged episodes of proximal supine reflux are noted.

symptoms. She continued to have asthma improvement. Her most recent clinic visit was October of 1996. She remains off corticosteroids and continues to do well. She has rare asthma exacerbations which are triggered by strong odors, tobacco smoke or an upper respiratory tract infection. She denies reflux symptoms almost two years after her fundoplication.

This case brings up several important points. First, despite frequent GERD symptoms, her physicians didn't realize that GERD was a trigger of JC's asthma. Furthermore, despite aggressive bronchodilators including prednisone, she had significant airflow obstruction. Second, acid suppressive therapy improved her asthma significantly, however, it took approximately three months for significant improvement. Pulmonary function tests also improved. Third, esophageal pH testing played a major role in JC's management. Her initial pH test showed abnormal amounts of proximal acid, a predictor of improved asthma outcome. She also had very high esophageal acid contact times leading to an endoscopy which showed esophagitis. Esophageal pH testing documented adequate acid suppression on her medical antireflux regimen and in the postoperative state. The final point is that she required a high dose of a proton pump inhibitor and a prokinetic agent to adequately control her GERD.

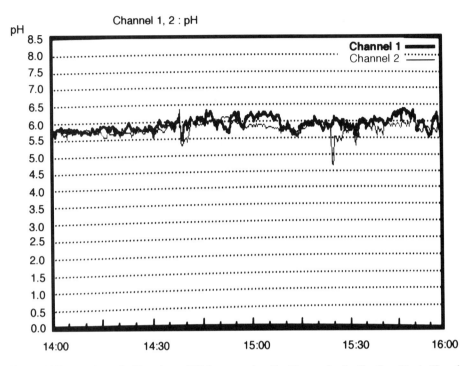

**Figure 13.7.** Esophageal pH tracing of JC four months after Toupet fundoplication. Dual pH probe placed in the distal esophagus (Channel 2, 5 cm above the LES) and the proximal esophagus (Channel 1, 2 cm below the UES). No significant reflux episodes are noted.

## CONCLUSIONS

Gastroesophageal reflux is a common trigger of asthma and needs to be considered in every asthmatic. Esophageal symptoms of GERD may be absent in approximately 25 percent of asthmatics with GERD. The mechanism of acid induced bronchoconstriction includes a vagally mediated reflex, however, microaspiration and heightened bronchial reactivity may also play a role. Aggressive anti-reflux therapy with proton pump inhibitors or anti-reflux surgery results in asthma improvement in 75 percent of asthmatics with GERD. Twenty four hour esophageal pH testing plays an important role in diagnosing GERD in asthmatics, especially in those who deny esophageal symptoms of GERD.

## REFERENCES

1. Osler WB: *The Principles and Practice of Medicine*, 8th ed. New York, D. Appleton and Co., 1912, p628–631.
2. Harding SM, Richter JE, Guzzo MR, et al: Asthma and gastroesophageal reflux: Acid suppressive therapy improves asthma outcome. *Am J Med* 100:395–405, 1996.

3. Irwin RS, Curley FJ, French CL: Difficult to control asthma. Contributing factors and outcome of a system management protocol. *Chest* 103:1662–1669, 1993.

4. Larrain A, Carrasco E, Galleguillos F, et al.: Medical and surgical treatment of nonallergic asthma associated with gastroesophageal reflux. *Chest* 99:1330–1335, 1991.

5. Sontag S, O'Connell S, Khandelwal S, et al.: Anti-reflux surgery in asthmatics with reflux (GER) improves pulmonary symptoms and function. *Gastroenterology* 98:A128, 1990 (Abstract).

6. Harding SM, Richter JE: Gastroesophageal reflux disease and asthma. *Semin Gastrointest Dis* 3:139–150, 1992.

7. Field SK, Underwood M, Brant R: Prevalence of gastroesophageal reflux symptoms in asthma. *Chest* 109:316–322, 1996.

8. Kjellen G, Brundin A, Tibbling L, et al.: Oesophageal function in asthmatics. *Eur J Respir Dis* 62:87–94, 1981.

9. Sontag SJ, Schnell TG, Miller TQ, et al.: Prevalence of oesophagitis in asthmatics. *Gut* 33:872–876, 1992.

10. Sontag SJ, O'Connell S, Khandelwal S, et al.: Most asthmatics have gastroesophageal reflux with or without bronchodilator therapy. *Gastroenterology* 99:613–620, 1990.

11. Mansfield LE, Stein MR: Gastroesophageal reflux and asthma: A possible reflex mechanism. *Ann Allergy* 41:224–226, 1978.

12. Mansfield LE, Hameister HH, Spaulding MS, et al.: The role of the vagus nerve in airway narrowing caused by intraesophageal hydrochloric acid provocation and esophageal distention. *Ann Allergy* 47:431–434, 1981.

13. Wright RA, Miller SA, Corsello BF: Acid-induced esophagobronchial cardiac reflexes in humans. *Gastroenterology* 99:71–73, 1990.

14. Tan WC, Martin RJ, Pandy R, et al.: Effects of spontaneous and stimulated gastroesophageal reflux on sleeping asthmatics. *Am Rev Respir Dis* 141:1394–1399.

15. Schan CA, Harding SM, Haile JM, et al.: Gastroesophageal reflux induced bronchoconstriction: An intraesophageal acid infusion study using state-of-the-art technology. *Chest* 106:731–737, 1994.

16. Harding SM, Schan CA, Guzzo MR, et al.: Gastroesophageal reflux induced bronchoconstriction: Is microaspiration a factor? *Chest* 108:1220–1227, 1995.

17. Harding SM, Guzzo MR, Maples RV, et al.: Gastroesophageal reflux induced bronchoconstriction: Vagolytic doses of atropine diminish airway responses to esophageal acid infusion. *Am J Respir Crit Care Med* 151:A589, 1995 (Abstract).

18. Herve P, Denjean A, Jian R, et al.: Intraesophageal perfusion of acid increase the bronchomotor response to methacholine and to isocapnic hyperventilation in asthmatic subjects. *Am Rev Respir Dis* 134:986–989, 1986.

19. Tuchman DN, Boyle JT, Pack AL, et al.: Comparison of airway responses following tracheal or esophageal acidification in the cat. *Gastroenterology* 87:872–881, 1984.

20. Jack CIA, Calverley PMA, Donnelly RJ, et al.: Simultaneous tracheal and oesophageal pH measurements in asthmatic patients with gastro-oesophageal reflux. *Thorax* 50:201–204, 1995.

21. DeVault KR, Castell DO; for the Practice Parameters Committee of the American College of Gastroenterology. Guidelines for the diagnosis and treatment of gastroesophageal reflux disease. *Arch Intern Med* 155:2165–2173, 1995.

22. American Gastroenterological Association. American Gastroenterological Association medical position statement: Guidelines on the use of esophageal pH recording. *Gastroenterology* 110:1981, 1996.

23. Kahrilas PJ, Quigley EMM: Clinical esophageal pH recording: A technical review for practice guideline development. *Gastroenterology* 110:1982–1996, 1996.

24. Schnatz PF, Castell JA, Castell DO: Pulmonary symptoms associated with gastroesophageal reflux. Use of ambulatory pH monitoring to diagnose and to direct therapy. *Am J Gastroenterol* 91:1715–8, 1996.

25. Sontag S, O'Connell S, Khandelwal S, et al.: Doe wheezing occur in association with an episode of gastroesophageal reflux? *Gastroenterology* 96:A482, 1989 (Abstract).

26. Kjellen G, Tibbling L, Wranne B: Effect of conservative treatment of oesophageal dysfunction on bronchial asthma. *Eur J Respir Dis* 62:190–197, 1981.

27. Goodall RJR, Earis JE, Cooper DN, et al.: Relationship between asthma and gastroesophageal reflux. *Thorax* 36:116–121, 1981.

28. Harper PC, Bergren A, Kaye MD: Anti-reflux treatment in asthma. Improvement in patients with associated gastroesophageal reflux. *Arch Intern Med* 147:56–60, 1987.

29. Nagel RA, Brown P, Perks WH: Ambulatory pH monitoring of gastro-oesophageal reflux in "morning dipper" asthmatics. *Br Med J* 297:1371–1373, 1988.

30. Ekstrom T, Lindgren BR, Tibbling L: Effects of ranitidine treatment on patients with asthma and a history of gastro-oesophageal reflux: A double blind cross over study. *Thorax* 44:19–23, 1989.

31. Depla AC, Bartelsman JF, Roos CM, et al.: Beneficial effect of omeprazole in a patient with severe bronchial asthma and gastroesophageal reflux. *Eur Respir J* 1:966–968, 1988.

32. Meier JH, McNally PR, Punja M, et al.: Does omeprazole (Prilosec) improve respiratory function in asthmatics with gastroesophageal reflux? A double-blind, placebo-controlled crossover study. *Dig Dis Sci* 39:2127–2133, 1994.

33. Ford GA, Oliver PS, Prior JS, et al.: Omeprazole in the treatment of asthmatics with nocturnal symptoms and gastro-oesophageal reflux: A placebo-controlled crossover study. *Postgrad Med J* 70:350–354, 1994.

34. Perrin-Fayolle M, Gormand F, Braillon G, et al.: Long-term results of surgical treatment for gastroesophageal reflux in asthmatic patients. *Chest* 96:40–45, 1989.

35. Tardif C, Nouvee G, Denis P, et al.: Surgical treatment of gastroesophageal reflux in ten patients with severe asthma. *Respiration* 56:115–116, 1989.

36. Overholt RH, Ashraf MM: Esophageal reflux as a trigger in asthma. *NY State J Med* 66:3030–3032, 1966.

37. Kennedy JH: "Silent" gastroesophageal reflux. *Dis Chest* 42:42–45, 1962.

38. Richter JE, Castell DO: Gastroesophageal reflux: Pathogenesis, diagnosis and therapy. *Ann Intern Med* 97:93–103, 1982.

39. Stein MR, Towner TG, Weber RW, et al.: The effect of theophylline on the lower esophageal sphincter pressure. *Ann Allergy* 45:238–241, 1980.

40. Kahrilas PJ: Gastroesophageal reflux disease. *JAMA* 276:983–988, 1996.

41. Maton PN: Omeprazole. *New Engl J Med* 324:965–975, 1991.

42. Klinkenberg-Knol EC, Festen HP, Jansen JB, et al.: Long-term treatment with omeprazole for refractory reflux esophagitis: Efficacy and safety. *Ann Intern Med* 121:161–167, 1994.

43. DeMeester TR, Bonavina L, Albertucci M: Nissen fundoplication for gastroesophageal reflux disease. Evaluation of primary repair in 100 consecutive patients. *Ann Surg* 204:9–20, 1986.

44. DeMeester TR, Bonavina L, Iascone C, et al.: Chronic respiratory symptoms and occult gastroesophageal reflux: A prospective clinical study and results of surgical therapy. *Ann Surg* 211:337–345, 1990.

45. Hinder RA, Filipi CJ, Wetscher A, et al.: Laproscopic Nissen fundoplication is an effective treatment for gastroesophageal reflux disease. *Ann Surg* 220:472–483, 1994.

# 14

# The Refractory Patient

Edgar Achkar, M.D.

Many therapeutic options are available for patients in whom the diagnosis of gastro-esophageal reflux disease has been objectively established. However, some patients fail to respond to the most aggressive forms of antireflux treatment. It is very difficult to know the exact proportion of patients who are refractory to treatment. The lack of response is dependent upon the diagnostic methods, patient selection, type of medical practice, and methods of follow-up. The number of patients failing treatment in clinical practice is not known. However, in spite of aggressive treatment using high doses of proton pump inhibitors about 12–20% of patients remain resistant to acid suppression.[1–3]

Before attempting to analyze the reasons for treatment failure, it is important to define what is meant by a refractory patient. A useful working definition refers to a patient who has made a reasonable attempt to make behavioral changes, who has been treated with high doses of a known effective agent for gastroesophageal reflux disease and who continues to complain of significant symptoms. The patient not responding because of a complication such as a stricture requiring frequent and multiple dilatations or the patient with dysplasia arising in Barrett's esophagus or other complications will not be considered here.

When faced with a patient who has received a long trial of potent antireflux treatment, several questions should be asked:

1. Is the patient compliant?
2. Is there is an underlying disorder aggravating GERD?
3. Has the right diagnosis been made?
4. Is the treatment administered sufficient?
5. Are the symptoms related to another form of esophagitis?
6. Is the patient resistant to gastric acid suppression?

## COMPLIANCE

A patient considered refractory to treatment is most often advised to take a proton pump inhibitor drug twice a day. The cost of such therapy is often overlooked by those who prescribe it. Yet, the cost drives many patients to avoid treatment entirely, decrease the dose or avoid taking the medication on a daily basis. Therefore the first step is to question the patient, the spouse, or another close member of the family about medication usage. At the same time, an attempt should be made to evaluate whether the patient has a good understanding of the factors leading to and aggravating gastroesophageal reflux. Smoking habits, analgesic use, weight changes, and late ingestion of meals should be reviewed to judge the patient's total compliance to treatment. This evaluation is extremely important before considering the patient as a "treatment failure" or embarking upon a more aggressive approach such as antireflux surgery.

# UNDERLYING MOTOR DISORDERS

### Case Study #1

A 40 year old man with insulin dependent diabetes was referred for intractable symptoms of gastroesophageal reflux disease. He gave a two year history of daily heartburn, frequent belching, occasional nausea, and regurgitation of acid material particularly when bending over. He also complained of abdominal distention after meals. An endoscopic examination had revealed no esophagitis. He was treated with omeprazole and cisapride with little benefit. Physical examination was unremarkable. An endoscopic examination showed no esophagitis, a small inflammatory polyp at the esophagogastric junction and some retained material in the stomach. A 24-hour pH test showed a slight increase in acid exposure in the upright position and there was no correlation between the episodes of heartburn and the decrease in pH (Figure 14.1). Esophageal motility test revealed incomplete relaxation of the lower esophageal sphincter and total absence of peristalsis. The diagnosis of achalasia was made and the patient treated accordingly.

The diagnosis of achalasia was delayed in this patient because of the severity of retrosternal burning and the minimal dysphagia he reported. Heartburn is not unusual in achalasia. It has been reported in as many as 40 percent of patients.[4,5] Studies including

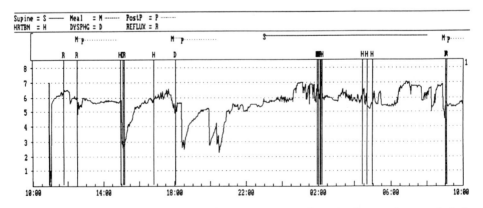

**Figure 14.1.** No significant reflux on pH study. Heartburn was due to stasis secondary to achalasia.

pH monitoring have shown abnormal acid exposure in 20–45 percent of untreated achalasia patients.[6,7] The most likely explanation for the presence of increased amounts of acid in the esophagus, however is the production of lactate secondary to food stasis rather than reflux of gastric acid.[8]

Patients who have a classic symptom complex of GERD and do not complain of dysphagia are evaluated by X-rays or endoscopy and placed on medical treatment if no structural lesions are detected. When the patient returns with continued symptoms, and the medical treatment has been increased without any additional response, serious consideration should be given to a motor disorder accounting for the lack of response. Patients, particularly women, should be questioned about Raynaud's phenomenon. A physical examination if not done earlier, should be performed to search for telangiectasias, sclerodactyly, and thickening of the skin on the arms or legs. Esophageal manometry may be indicated to rule out a specific motor disorder such as scleroderma or other connective tissue disease. Testing may also reveal nonspecific motor abnormalities such as low amplitude contractions or failed peristalsis, findings which may account for poor esophageal clearance and therefore lack of response to medical treatment. The importance of peristaltic dysfunction in patients with reflux disease and its potential contribution to increased esophageal exposure to refluxed acid was shown by Kahrilas et al.[9] In a study of patients with various degrees of reflux esophagitis and controls, they found that failed primary peristalsis or decreased lower esophageal contraction amplitude was present in 25 percent of patients with mild esophagitis and 48 percent of patients with severe esophagitis.

# DIAGNOSTIC ACCURACY

The classic patient with GERD complains of heartburn occurring day and night, responding partially to antacids, and aggravated by certain foods and certain positions. Other patients are treated because of less classic symptoms such as chest pain, regurgitation, waterbrash, and dyspepsia. When treatment appears to have failed, time should be taken to review the history in detail as if this was the first encounter with the patient. At that point, many patients have learned to use medical terms such as reflux, acidity, hiatal hernia, and so on. I make an effort to ask them to describe their symptoms and ignore the terms they have learned from their physicians. When doing this, one finds very often that what was thought to be a classic history for GERD may not be so. At any rate, this is a good time to assess objectively whether the symptoms are related to excessive gastric acid exposure. A 24-hour pH test, after treatment has been discontinued, is the most useful approach.

### Case Study #2

A 34 year old patient was referred for evaluation prior to antireflux surgery because of continued symptoms after prolonged medical treatment. The patient was diagnosed over 18 months earlier; endoscopic examination revealed a small hiatal hernia but no esophagitis. She had no dysphagia, denied the use of aspirin or NSAIDs and had no evidence of any systemic disease. She had already received an $H_2$ receptor antagonist twice a day and, when she continued to complain, the dose was doubled. After several weeks she was started on omeprazole 20 mg a day. She continued to report no improvement and was then started on 20 mg of omeprazole

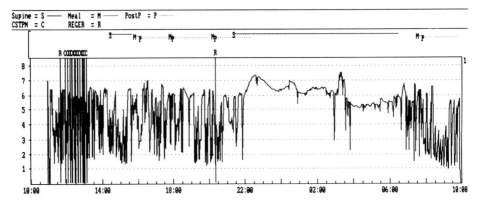

**Figure 14.2.** Tracing show multiple "reflux" episodes during daytime with a stable normal pH through the night. pH fluctuations are brief and associated with cough and belching.

twice a day. Endoscopic examination was repeated showing no evidence of esophagitis, stricture, or other lesions. The physical examination was unremarkable. Careful questioning revealed that her symptoms consisted of retrosternal pressure occurring throughout the day, unrelated to meals, rarely nocturnal and accompanied by a sour taste in the mouth and frequent belching and regurgitation. Esophageal manometry showed a normal lower esophageal sphincter pressure and normal peristalsis. She was asked to discontinue omeprazole for seven days and 24-hour esophageal pH monitoring was performed. The test showed abnormal values during the day and no acid exposure at night. Careful analysis of the record revealed that the drops in pH were all very brief and often associated with cough, chest pain or belching (Figure 14.2). The findings were explained to the patient, she was counseled to avoid belching and her dietary habits were reviewed. She was discouraged from seeking surgical treatment.

This case illustrates the importance of symptom assessment and the value of pH monitoring in the evaluation of the so-called "refractory" patient.

## TREATMENT ADEQUACY

The mainstay of medical treatment for GERD remains acid suppression. It is well known now that different patients require different degrees of acid suppression. Before the advent of proton pump inhibitors, this observation was made clinically based on the fact that some patients required only two doses of histamine receptor antagonist a day whereas others required double or triple the regular dosing regimen. Collen and coauthors[10] compared a group of nonrespondents to a similar group of patients responding to ranitidine 150 mg twice a day. They found that the nonresponders had a higher degree of gastric acidity measured by gastric analysis than the group of responders. They then treated the refractory group with twice the original dose of ranitidine (300 mg B.I.D.) producing a significant improvement in symptoms. Although, it is valid to escalate acid suppressive treatment in patients who do not respond, the correlation between gastric acidity and severity of symptoms is not established. For example, Hirschowitz[11] measured basal and stimulated acid and pepsin outputs in 155 patients with endoscopically proven esophagitis and compared them to 508 controls. Neither the composition of gastric juice nor the basal or stimulated gastric acid or pepsin output correlated with the presence or the severity of esophagitis.

An extension of the concept of "the more the acid, the less the response" is represented by the patient with a hypersecretory condition. It is reasonable to think of the Zollinger-Ellison syndrome in patients not responding to treatment.[12] However, acid reflux resulting from a gastrinoma is usually quite severe and manifests itself by an advanced degree of esophagitis and the presence of peptic ulcer disease on endoscopy. Although a serum gastrin is often obtained in the patient not responding to medical therapy, in the absence of other features of gastrinoma, it is rarely useful.

# DIFFERENTIAL DIAGNOSIS

We associate reflux esophagitis with a typical symptom complex of heartburn, regurgitation, and waterbrash. When esophagitis is present and the patient fails to respond to aggressive acid suppression, the question arises as to the possibility of another noxious agent causing the esophagitis. Infectious esophagitis such as candidiasis, CMV and other infections occurring mostly in immuno-compromised patients have a totally different clinical presentation (odynophagia, dysphagia) and are rarely confused with acid reflux esophagitis.[13,14] However, over the years, another type of refluxate has been implicated in the etiology of reflux esophagitis. Some have suggested that "alkaline" gastroesophageal reflux may play a significant role in esophagitis and even Barrett's esophagus and adenocarcinoma of the esophagus.[15] Unlike acid reflux, alkaline exposure in the esophagus is difficult to measure. Recently a spectrophotometer detecting the presence of bilirubin was developed. Vaezi et al.[16] showed that the Bilitec measures adequately the presence of bile although it may be underestimated in the presence of acid. Clinical studies in postgastrectomy patients as well as in patients with GERD showed that bile reflux tended to be higher in patients with esophagitis but its presence was always associated with marked acid exposure.[17] Therefore the conventional wisdom of treating reflux esophagitis with acid suppression remains sound.

# RESISTANCE TO GASTRIC ACID SUPPRESSION

## Case Study #3

A 42 year old man presented with a five year history of retrosternal burning, frequent regurgitation, and waterbrash. His symptoms were intermittent at first but became gradually worse, waking him up at least two to three times a week. At first antacids provided some relief but he had to start using $H_2$ receptor antagonists over the counter. Recently, he was given a prescription for famotidine 20 mg twice a day but relief was temporary and partial. An endoscopic examination revealed erosive esophagitis, no evidence of Barrett's esophagus and a normal stomach and duodenum. A 24-hour pH test showed marked acid exposure during the day as well as at night with prolonged periods of acid exposure and good correlation with symptoms (Figure 14.3). The patient was started on omeprazole 20 mg a day and advised about lifestyle changes. He did not smoke and used alcohol rarely. After four weeks, he con-

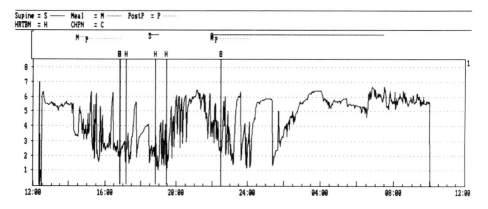

**Figure 14.3.** Ambulatory pH monitoring showing marked reflux: time pH < 4 is 26% upright and 22 percent supine.

tinued to report being awakened at night, but now only twice a week and his heartburn continued to occur during the day although less frequently than before. The dose of omeprazole was doubled and, after an additional four weeks, he reported some improvement but continued to complain of daily heartburn. A repeat 24-hour pH test while the patient was taking 20 mg of omeprazole twice a day revealed continued abnormal parameters although acid exposure was less than on the prior study (Figure 14.4). Endoscopic examination revealed healing of the erosive esophagitis. Antireflux surgery was discussed with the patient and a laparoscopic fundoplication was performed. Postoperatively, the patient was asymptomatic, ate a normal diet, and was able to stop all medications. A pH study revealed marked improvement (Figure 14.5).

This case illustrates that a certain subset of patients do not respond to aggressive acid suppression and should be considered for surgical treatment.

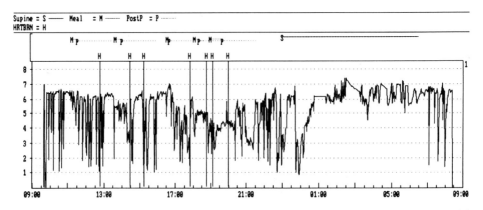

**Figure 14.4.** Significant reduction in acid exposure with omeprazole 20 mg BID: pH time < 4 is 9.7 percent upright and 8.1 percent supine.

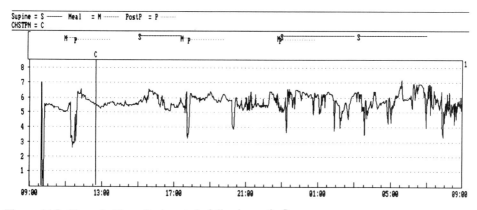

**Figure 14.5.** Normal pH monitoring study following antireflux surgery.

Suppression of gastric acid secretion is accomplished in a good proportion of patients with $H_2$ receptor antagonists. Proton pump inhibitors produce an even higher degree of suppression in nonresponsive patients.[18] However, as already pointed out, some patients continue to be symptomatic. Kuo and Castell[19] recently found that administering omeprazole in divided doses of 20 mg twice a day achieved more gastric acid suppression than 40 mg given in one dose. However the study was performed in normal, healthy volunteers and, in spite of this aggressive regimen, acid reflux continued to occur. Therefore, the relevance of these findings for treating patients with reflux disease is not clear. Katzka et al. advocate using 24 hour pH monitoring to correlate reflux episodes with persistent symptoms in patients not responding to conventional doses of omeprazole.[20] Leitel et al.[3] tried to characterize "proton pump inhibitor resistance" objectively. They studied 88 patients with reflux disease whose symptoms were resistant to conventional treatment. The authors placed these patients on 20 mg of omeprazole twice a day while obtaining ambulatory intragastric and distal esophageal pH monitoring. These patients were compared to random controls as well as patients with reflux disease who had responded to treatment. The authors defined "omeprazole failure" or "abnormal acid secretion under treatment" as a gastric pH $< 4$ for more than 50 percent of the time. On B.I.D. omeprazole therapy, GERD controls and normal individuals showed intragastric pH of less than 4 for 30 percent of the time as compared to 80% for patients not responding to omeprazole. When the dose of omeprazole was increased to 80 mg a day, omeprazole failure patients showed a decrease of pH time exposure from 74.3 to 32.8 percent suggesting that the phenomenon may be dose dependent.

## SUMMARY

Prolonged pH monitoring is extremely useful in the evaluation of patients refractory to treatment. The procedure helps to establish the right diagnosis, provides objective evidence of acid reflux and confirms that, in some cases, the most aggressive attempt at acid suppression is not sufficient. The results of pH monitoring can be used to alter medical therapy and the test constitutes a significant element for the indication to antireflux surgery.

# REFERENCES

1. Hatlebakk JG, Berstad A, Carling L, et al: Lansoprazole versus omeprazole in short term treatment of reflux esophagitis: Results of a Scandinavian multicenter trial. *Gastroenterology* 102:A80 (abstract), 1992.
2. Laursen LS, Bondesen S, Hansen J, et al: Omeprazole 20 mg or 40 mg for the treatment of gastroesophageal reflux disease? A double blind comparative study. *Gastroenterology* 102:A110 (abstract), 1992.
3. Leite LP, Johnston BT, Just RJ, Castell DO: Persistent acid secretion during omeprazole therapy: A study of gastric acid profiles in patients demonstrating failure of omeprazole therapy. *Am J Gastroenterol* 91:1527–1531, 1996.
4. Howard PJ, Maher L, Pryde A, Cameron EW, Heading RC: Five year prospective study of the incidence, clinical features, and diagnosis of achalasia in Edinburgh. *Gut* 33(8):1011–1015, 1992.
5. Goldenberg SP, Vos C, Burrell, M, Traube M: Achalasia and hiatal hernia. *Dig Dis Sci* 37:528–531, 1992.
6. Smart HL, Foster PN, Evans DF, Slevin B, Atkinson M: Twenty four hour oesophageal acidity in achalasia before and after pneumatic dilatation. *Gut* 28:883–887, 1987.
7. Ferraro P, Perrault L, Emond C, Filion R, B3aucham G: Preoperative 24 hour pH monitoring in achalasia patients. *Dis Esophagus* 8:2000–2004, 1985.
8. Burke CA, Achkar E, Falk GW: The effect of pneumatic dilation (PD) on gastroesophageal reflux. *Am J Gastroenterol* 89:1615;1994.
9. Kahrilas PJ, Dodds WJ, Hogan WJ, Kern M, et al: Esophageal peristaltic dysfunction in peptic esophagitis. *Gastroenterology* 91:897–904, 1986.
10. Collen MJ, Lewis JH, Benjamin SB: Gastric acid hypersecretion in refractory gastroesophageal reflux disease. *Gastroenterology* 98:654–661, 1990.
11. Hirschowitz BI. A critical analysis, with appropriate controls, of gastric acid and pepsin secretion in clinical esophagitis. *Gastroenterology* 101:1149–1158, 1991.
12. Richter JE, Pandol SJ, Castell DO, McCarthy DM: Gastroesophageal reflux disease in the Zollinger-Ellison Syndrome. *Ann Int Med* 95:37–43, 1981.
13. Wilcox CM, Schwartz DA, Clark WS: Esophageal ulceration in human immunodeficiency virus infection. *Ann Intern Med* 122:143–149, 1995.
14. Wilcox CM, Diehl DL, Cello JP, et al: Cytomegalovirus esophagitis in patients with AIDS. *Ann Intern Med* 113:589–593, 1990.
15. Attwood SEA, Ball CS, Barlow AP, Jenkinson L, et al: Role of intragastric and intraoesophageal alkalinisation in the genesis of complications in Barrett's columnar lined lower oesophagus. *Gut* 34:11–15, 1993.
16. Vaezi MF, Lacamera RG, Richter JE: Validation studies of Bilitec 2000: an ambulatory duodenogastric reflux monitoring system. *Am J Physiol* 267 (Gastrointest Liver Physiol 30):G1050–G1057, 1994.
17. Champion G, Richter JE, Vaezi MF, et al: Duodenogastroesophageal reflux: relationship to pH and importance in Barrett's esophagus. *Gastroenterology* 107:747–754, 1994.
18. Klinkenberg-Knol EC, Festen HPM, Jansen JBMJ, et al. Long-term treatment with omeprazole for refractory reflux esophagitis: Efficacy and Safety. *Ann Int Med* 121:161–167, 1994.
19. Kuo B, Castell DO. Optimal dosing of omeprazole 40 mg daily: Effects on gastric and esophageal pH and serum gastrin in healthy controls. *Am J Gastroenterol* 91:1532–1538, 1996.
20. Katzka DA, Paoletti V, Leite L, Castell DO. Prolonged ambulatory pH monitoring in patients with persistent gastroesophageal reflux disease symptoms: Testing while on therapy identifies the need for more aggressive anti-reflux therapy. *Am J Gastroenterol* 91:2110–2113, 1996.

# 15

# Pre- and Postoperative Use of Ambulatory 24 Hour pH Monitoring and Manometry

**Ross M. Bremner, M.D.**
**Tom R. DeMeester, M.D.**
**Hubert J. Stein, M.D.**

Gastroesophageal reflux is a common disease that accounts for approximately 75 percent of esophageal pathology. The prevalence of gastroesophageal reflux disease in the United States is assumed to be as high as 0.36 percent, or 360 affected individuals per 100,000 population.[1] Prior to recommending surgical therapy for gastroesophageal reflux disease, an objective diagnosis of the presence of the disease is necessary, and the cause of increased esophageal exposure to gastric juice needs to be established.[2]

Until esophageal exposure to gastric juice could be measured by prolonged monitoring of esophageal pH, an objective documentation of the disease was difficult. This is because symptoms or the finding of endoscopic or histologic esophageal mucosal injury is an unreliable guide to the presence of gastroesophageal reflux disease. Only after introduction of 24 hour esophageal pH monitoring can the basic pathophysiologic abnormality of gastroesophageal reflux disease, i.e., increased esophageal exposure to gastric juice, be measured.[3-5] Extensive clinical experience and validation studies have confirmed that 24 hour esophageal pH monitoring, when analyzed according to the principles outlined in Chapter 8, is highly reproducible and has the highest sensitivity and specificity for the detection of gastroesophageal reflux disease. *It does not, however, determine the reason for the increased exposure.*

## PHYSIOLOGIC BASIS OF SURGICAL THERAPY FOR GASTROESOPHAGEAL REFLUX DISEASE

There are three known causes of increased esophageal exposure to gastric juice in patients with gastroesophageal reflux disease (Figure 15.1).[1] The first is a mechanically de-

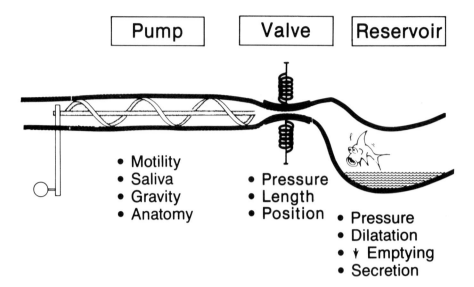

**Figure 15.1.** Mechanical model of the esophagus as a propulsive pump, the lower esophageal sphincter as a valve, and the stomach as a reservoir. Esophageal clearance of refluxed gastric juice is determined by the esophageal motor activity, salivation, gravity, and presence of a hiatus hernia. The competency of the lower esophageal sphincter depends on its pressure, overall length, and length exposed to abdominal pressure. Gastric function abnormalities that cause gastroesophageal reflux include increased intragastric pressure, gastric dilation, decreased emptying rate, and increased gastric acid secretion.

fective lower esophageal sphincter. This accounts for about 50 to 60 percent of gastroesophageal reflux disease.[6] The identification of a mechanically defective sphincter is important because it is the one etiology that antireflux surgery is designed to correct. The other two causes of increased esophageal acid exposure are inefficient esophageal clearance of refluxed gastric juice and abnormalities of the gastric reservoir that augment physiologic reflux. The factors important in esophageal clearance are gravity, esophageal motor activity, salivation, and anchoring of the distal esophagus in the abdomen. Gastric abnormalities that increase esophageal exposure to gastric juice are increased gastric pressure, excessive gastric dilation, increased gastric acid secretion, and/or a persistent gastric reservoir secondary to delayed gastric emptying. When increased esophageal acid exposure is caused by ineffective esophageal clearance or gastric abnormalities, an antireflux may be procedure may not be indicated because of the potential to induce dysphagia and gastric symptoms.[7]

In humans, the resistance to gastroesophageal reflux provided by the lower esophageal sphincter is related to the sphincter pressure, exposed to the positive pressure environment of the abdomen. From a clinical perspective, mechanical failure of the lower esophageal sphincter is diagnosed by esophageal sphincter manometry. Table 15.1 shows the normal values of the lower esophageal sphincter pressure, overall length, and abdominal length from 50 healthy volunteers. The manometric definition of a mechanically defective lower esophageal sphincter was developed by comparing the frequency distribu-

TABLE 15.1. NORMAL MANOMETRIC VALUES OF THE LOWER ESOPHAGEAL SPHINCTER (N = 50)

| | Mean | Mean −2 Standard Deviations | Mean +2 Standard Deviations |
|---|---|---|---|
| Pressure (mm Hg) | 13.8 ± 4.6 | 4.6 | 23.0 |
| Overall length (cm) | 3.7 ± 0.8 | 2.1 | 5y.3 |
| Abdominal length (cm) | 2.2 ± 0.8 | 0.6 | 3.8 |

| | | Percentile | |
|---|---|---|---|
| | Median | (2.5) | (97.5) |
| Pressure (mm Hg) | 13.0 | 5.8 | 27.7 |
| Overall length (cm) | 3.6 | 2.1 | 5.6 |
| Abdominal length (cm) | 2.0 | 0.9 | 4.7 |

tion of the values of these sphincter components in the 50 normal subjects with a population of 622 patients with symptoms of gastroesophageal reflux disease, 324 of whom had increased esophageal exposure to gastric juice on 24 hour esophageal pH monitoring.[6] On the basis of these data, we have defined a mechanically defective lower esophageal sphincter as having one or more of the following a mean lower esophageal sphincter pressure of less than 6 mm Hg, a mean lower esophageal sphincter length exposed to the positive pressure environment in the abdomen of less than 1 cm, and a mean overall sphincter length of less than 2 cm. These values are below the 2.5 percentile for sphincter pressure and overall length and below the fifth percentile for abdominal length in normal subjects. The probability of increased esophageal exposure to gastric juice is 69 to 76 percent if one of these components of the sphincter is abnormal, 65 to 88 percent if two components are abnormal, and 92 percent if all three are abnormal.[6] This indicated that a failure of one or two of these components may be compensated for by the clearance function of the esophageal body. Failure of all three components inevitably leads to increased esophageal exposure to gastric juice.

Figure 15.2 shows that the development of complications of gastroesophageal reflux disease is related to the presence of a mechanically defective lower esophageal sphincter.[8] This indicates that a mechanically defective sphincter is the major factor in the pathogenesis of complications. The observation that a mechanically defective sphincter also occurs in about 28 percent of patients who do not have a complication of increased esophageal exposure to gastric juice suggests that the defect in the sphincter is primary and not the result of inflammation or tissue damage.

There has been considerable interest in so-called TLESR's or transient lower esophageal sphincter relaxations as an explanation for gastroesophageal reflux, particularly in patients with mechanically normal sphincters. We have recently shown this phenomenon to be associated with a progressive "taking up" of the sphincter by the fundus as a consequence of gastric distention (Figure 15.3).[9] The LES length is reduced down to a critical value at which competency is lost due to mechanical factors that pull the sphincter open. This explains the occurrence of post-prandial reflux and its gradual abatement as the ingested volume decreases with gastric emptying (Figure 15.4). Naturally, a sphincter

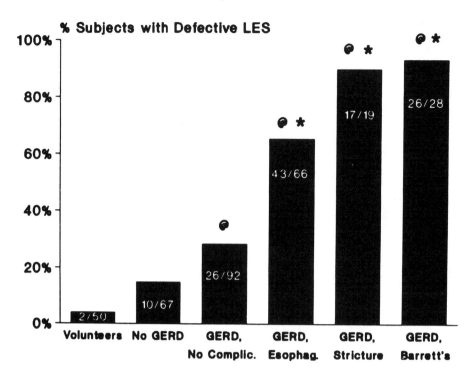

**Figure 15.2.** Prevalence of a mechanically defective sphincter in patients with gastroesophageal reflux disease (GERD) and no complications, esophagitis, stricture, or Barrett's esophagus.

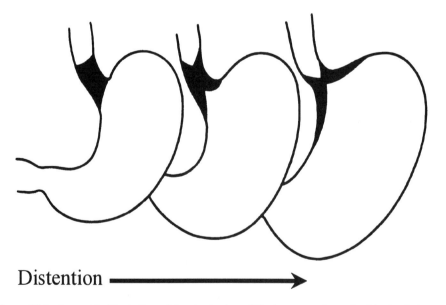

**Figure 15.3.** A graphic illustration of the shortening of the lower esophageal sphincter that occurs as the sphincter is "taken up" by the cardia as the stomach distends.

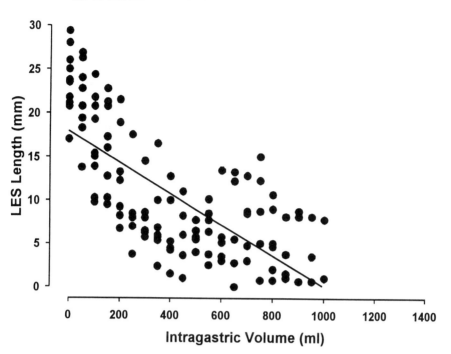

**Figure 15.4.** A graph illustrating the decrease in the length of the sphincter that occurs with increasing gastric volume.

with a shorter abdominal or overall length, or one with a lower resting pressure will be more prone to lose competency with even a minor degree of unfolding. In this situation, the use of surgery is based on our observation in the animal model showing that the Nissen fundoplication, although returning the competency of a failed sphincter to normal, also acts to prevent shortening of the sphincter, thereby preventing reflux. Consequently, we no longer require a mechanically defective sphincter to be present prior to antireflux surgery. We now operate on patients who have normal sphincters but evidence of increased esophageal acid exposure on pH monitoring. For the most part, these are patients with early GERD who are amenable to a fundoplication done laparoscopically.

# INDICATIONS FOR ANTIREFLUX SURGERY BASED ON ESOPHAGEAL FUNCTION TESTS

Before an antireflux procedure is considered in a patient with gastroesophageal reflux disease, it is necessary to confirm that the patient's symptoms are attributable to increased esophageal exposure to gastric contents secondary to a mechanically defective lower esophageal sphincter. This requires performing esophageal function studies, i.e., 24 hour esophageal pH monitoring and esophageal manometry. As outlined in the algorithm in Figure 15.5, esophageal function studies should be done if a patient has persistent symptoms or unimproved esophageal mucosal injury after 8 to 12 weeks of intensive acid suppression

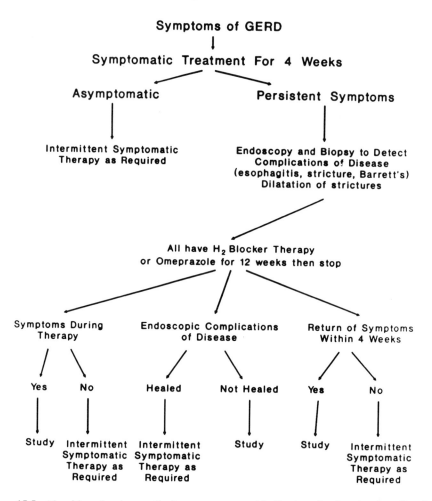

**Figure 15.5.** Algorithm showing medical management and indications for functional studies (i.e., 24 hour pH monitoring and manometry) in patients with symptoms of gastroesophageal reflux disease (GERD).

therapy. Patients who respond to a course of medical therapy but have recurrence of symptoms within 4 weeks after cessation of therapy should also be studied; since they are prone to drug dependency. Costantini and coworkers have recently shown that patients with a mechanically defective sphincter defined as above, are destined to become dependent on medical therapy to control symptoms.[10] We recently have identified risk factors, which, if present, prompt early surgical intervention. The risk factors that portend to a poor outcome with medical treatment are: 1) mechanically defective valve, 2) increased esophageal exposure to bile, 3) supine reflux, 4) severe esophagitis on initial endoscopy, 5) Barrett's esophagus, and 6) young age. A strong case can be made that these patients respond poorly to medical management, and early surgical intervention provides a better outcome.

The two criteria that should be present in order for a surgeon to proceed with an antireflux procedure in a patient managed as outlined in Figure 15.5 are

1. persistent or recurrent symptoms and/or complications after 8 to 12 weeks of intensive acid suppression therapy *and*
2. increased esophageal exposure to gastric juice documented by 24 hour esophageal pH monitoring.

If 24 hour esophageal pH monitoring is normal in a patient with unequivocal endoscopic esophagitis, the possibilities of alkaline, drug-induced, or retention esophagitis should be considered. If the sphincter is manometrically normal in a patient with increased esophageal exposure to gastric juice, he or she should be evaluated for an esophageal or gastric cause of increased acid exposure prior to surgical therapy. Patients with increased acid exposure and a mechanically defective sphincter who have no complications of the disease and whose control of symptoms requires longterm drug dependency should be given the option of surgery as a cost-effective alternative.[1]

If the patient's symptoms respond to medical therapy but endoscopic esophagitis persists, surgery should be performed. If an antireflux procedure is done before the development of complications, it will correct the mechanically defective sphincter, heal esophagitis, and prevent the formation of a stricture or Barrett's esophagus.[7] It is important to realize that even the new generation proton pump inhibitors may not completely stop reflux. Rather they change the pH of the gastric juice and reduce the volume of the gastric secretions. Because the mechanical defect persists, these agents may reduce, but do not eliminate reflux. There is recent evidence to suggest that damage may continue to the underlying esophageal mucosa in this situation.[8] Furthermore, patients with respiratory symptoms may continue to aspirate and, in fact, may be worse off since they have lost the warning signs of heartburn that reflux is occurring.

The development of a stricture in a patient with increased esophageal exposure to gastric juice on esophageal pH monitoring and a mechanically defective sphincter is usually associated with loss of esophageal contractility.[11] This represents a situation in which medical therapy has failed and the loss of esophageal function has occurred prior to the clinician's recommending a surgical antireflux procedure. Before surgery is performed, a malignant etiology of the stricture should be excluded and the stricture should be progressively dilated up to a No. 60 French bougie. When the stricture is fully dilated, the relief of dysphagia is evaluated, and esophageal manometry is performed to determine the adequacy of peristalsis in the distal esophagus. If dysphagia is relieved and the amplitude of esophageal contractions is adequate, an antireflux procedure should be performed; if the amplitude of esophageal contractions is poor, caution should be exercised in performing an antireflux procedure with a complete fundoplication, and a partial fundoplication should be considered. If the dysphagia persists and the amplitude of contractions is deficient, an esophageal resection and colon interposition should be considered.[12] In rare situations dysphagia persists in the presence of an adequate amplitude of esophageal contractions. In this situation we recommend combined monitoring of esophageal pH and motor activity over 24 hours to evaluate for an unrecognized motor disorder prior to proceeding with an antireflux procedure.[11] If after dilation the 24 hour esophageal pH record in a patient with a stricture is normal, the stricture is probably secondary to drug ingestion and dilation may be all that is needed.

Barrett's columnar-lined esophagus is almost always associated with increased esophageal exposure to gastric juice on 24 hour pH monitoring, a severe mechanical defect of the lower esophageal sphincter, and poor contractility of the esophageal body.[11,13] Patients with Barrett's esophagus are at risk of progression of the mucosal abnormality up the esophagus, formation of a stricture, hemorrhage from a Barrett's ulcer, and the development of an adenocarcinoma. A surgical antireflux procedure can arrest the progression of the disease, heal ulceration, and resolve strictures. An antireflux procedure may also reduce the degree of pleomorphism and dysplasia. If on mucosal biopsies severe dysplasia or intramucosal carcinoma in situ is found, an esophageal resection should be done.[13]

Chronic atypical symptoms of reflux, e.g., chest pain, chronic cough, recurrent pneumonias, episodes of nocturnal choking, waking up with gastric contents in the mouth, or soilage of the bed pillow, are also indications for esophageal function studies.[14,15] If 24 hour pH monitoring confirms the presence of increased esophageal acid exposure and manometry shows a mechanical defect of the lower esophageal sphincter and normal esophageal body function, an antireflux procedure is indicated with an expected good result. It is not unusual for these patients to have a nonspecific motor abnormality of the esophageal body, which tends to propel the refluxed material towards the pharynx. In some of these patients the motor abnormality disappears following reflux control by a surgical antireflux procedure. In others the motor disorder persists and may cause postoperative aspiration of swallowed saliva and food. Consequently, the results of an antireflux procedure in patients with a motor disorder of the esophageal body are variable.[14] In this situation, combined ambulatory monitoring of esophageal pH and motility over 24 hours shows promising initial results in identifying those patients who may benefit from an antireflux procedure in this situation.

Prior to proceeding with an antireflux procedure in a patient with increased esophageal exposure to gastric juice and a mechanically defective lower esophageal sphincter, the surgeon should specifically question the patient about complaints of epigastric pain, nausea, vomiting, and loss of appetite. In the past, these symptoms were accepted as part of the reflux syndrome, but we now realize that they can be caused by excessive duodenogastric reflux, which occurs in about 28 percent of patients with gastroesophageal reflux disease.[8] This problem is usually but not always seen in patients who have had previous upper gastrointestinal surgery. In such patients, correction of only the mechanically defective lower esophageal sphincter will result in a disgruntled individual who continues to complain of nausea and epigastric pain on eating. In these patients, 24 hour pH monitoring of the stomach may help to detect and quantitate duodenogastric reflux.[16] The diagnosis can also be documented with a $^{99m}$Tc-HIDA scan if excessive reflux of bile from the duodenum into the stomach can be demonstrated.[17] If surgery is necessary to control gastroesophageal reflux, and if severe duodenogastric reflux is present, consideration should also be given to performing a bile diversion procedure.[18]

# THE BENEFITS OF SIMULTANEOUS ESOPHAGEAL MONITORING OF PRESSURES AND pH

Additional information is offered by simultaneously recording pressures during ambulatory pH monitoring which help to interpret the pH study. These are 1) the relationship of respiratory symptoms such as coughing to reflux episodes, 2) interpretation of the effect of "napping" on reflux, 3) identification of periods of drinking acidic liquids as an artefactual cause of reflux, 4) detection of reflux induced motor abnormalities. An awareness of these patterns is important in the interpretation of pH studies.

There is an increased interest in the relationship between reflux disease and respiratory symptoms such as asthma and chronic cough.[19-22] Detecting an association between acid reflux and respiratory symptoms during standard pH testing depends on the patients documentation of symptoms, from which a correlation with the pH record is made. This method is limited by the compliance of the patient reporting symptoms, and an inability to appreciate that coughing can either cause, or be caused by the reflux event. Coughing episodes have a characteristic morphology as recorded by the ambulatory manometry catheter. The simultaneous, isobaric pressure increases have a steep upslope (dp/dt) which can be recognized visually and by the computer. This provides a means of identifying when a cough occurs and relating it to changes in pH during the same time interval. Figure 15.6 shows an episode of coughing preceding a reflux event. In this situation the reflux may have been precipitated by the increase in intra-abdominal pressure caused by coughing. The coughing episode was not caused by the reflux episode. On the other hand, when the coughing follows a reflux episode, the reflux may have precipitated the coughing (Figure 15.7). It is expected that the combination of pH and pressure monitoring in the future will allow the accurate documentation of coughing in association with reflux and consequently will enable a more rational approach to antireflux surgery in this situation.

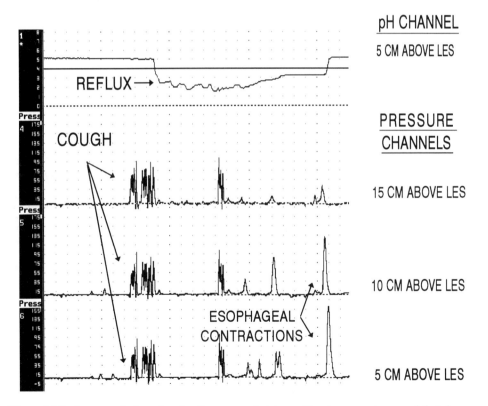

**Figure 15.6.** An example from the simultaneous ambulatory pH and manometry record that shows coughing preceding a reflux episode. In this situation the increase in intra-abdominal pressure from the coughing may have precipitated the reflux episode.

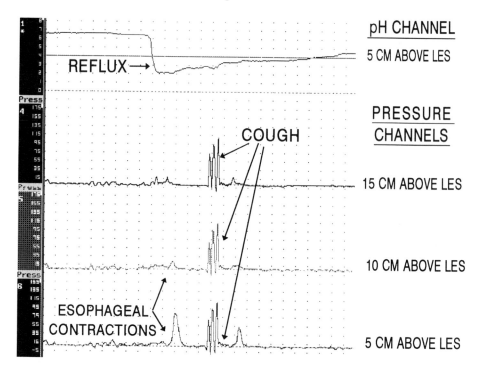

**Figure 15.7.** An example from the simultaneous ambulatory pH and manometry record that shows coughing following the onset of a reflux episode. In this situation the reflux may have precipitated coughing.

During sleep esophageal motility is quiescent because of a decreased swallowing rate. This provides an explanation for the longer reflux episodes that occur at night. It has been noted than many patients "doze" during the post-prandial period which has significant implications since reflux is more common after eating.[23-26] Ambulatory manometry provides a mechanism to identify sleep periods because of the remarkable decrease in esophageal activity. Figure 15.8 shows a compressed record of the simultaneous pH and motility study. Seen in isolation, the pH record (top channel) reveals a long reflux episode ($\pm70$ minutes) in the afternoon. The ambulatory manometry record shows that this post-prandial episode is related to sleep as the activity of the esophagus is dramatically decreased. This was later confirmed by direct questioning of the patient who remembered falling asleep in an armchair after lunch. The long episode is consequently a result of poor arousal by the reflux event and no initiation of swallowing to clear the episode. This could increase the potential for aspiration. If the pH study had been performed without simultaneous pressure monitoring, the lack of esophageal body function would not have been appreciated.

Some carbonated beverages such as cola have an acidic pH. Drinking these liquids causes a drop in intraesophageal pH which could be interpreted as a reflux event if the patient does not follow the dietary instructions. An example of this is shown in Figure 15.9. Seen in isolation, the pH record would be interpreted as a reflux episode. Even the addition of esophageal manometry (lower three channels) does not dispute this as the contractions of the esophageal body appear to be a normal clearing peristaltic wave in

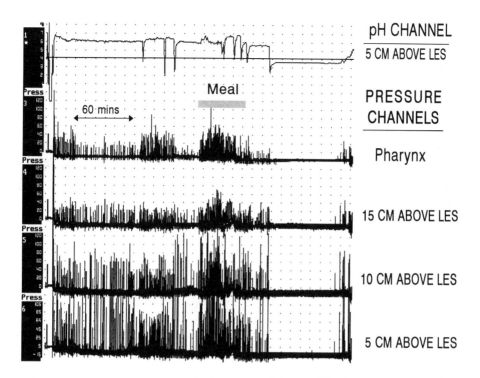

**Figure 15.8.** An example from the simultaneous ambulatory pH and manometry record compressed to show 300 minutes of the 24-hour study. Shortly after lunch a long reflux episode is associated with a marked decrease in esophageal activity during an afternoon nap.

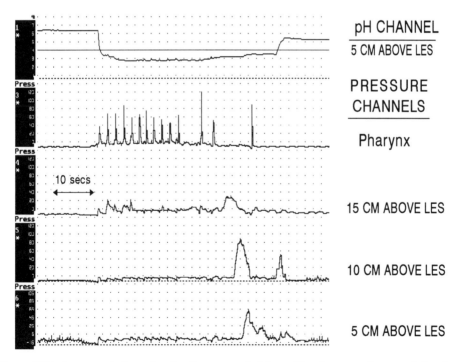

**Figure 15.9.** An example from the simultaneous ambulatory pH and manometry record showing an artifactual reflux episode. The drop in esophageal pH is related to the drinking of a carbonated, acidic soda, and not to gastroesophageal reflux.

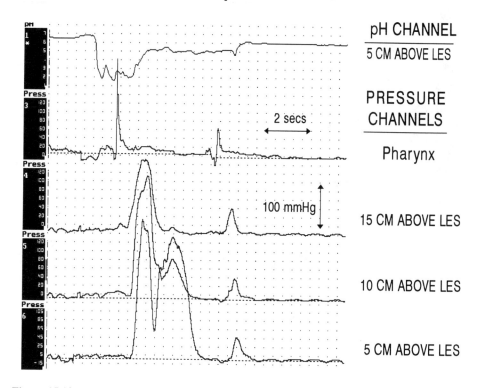

**Figure 15.10.** An example from the simultaneous ambulatory pH and manometry record showing "reflux induced spasm". Note the normal contractions after the reflux episode.

response to the reflux event. The addition of the pharyngeal channel, however, shows that this is not a reflux event at all. Rather, the rapid swallowing indicates drinking. The drop in esophageal pH occurs immediately after the first swallow. This is confirmed by the slight elevation in the baseline seen in the esophageal channels, as liquid fills the lumen. The end of the train of swallows is followed by a peristaltic wave, providing, in this instance, a good example of deglutitive inhibition in the physiological setting. The patient later admitted to drinking cola between meals on 4 or 5 different occasions during the study.

Ambulatory motility and pH monitoring in patients with reflux disease is also able to document reflux induced motor abnormalities. We have seen the occasional patient presenting with non-cardiac chest-pain and normal stationary manometry. Their recorded contractions induced by reflux episodes look like esophageal spasm (Figure 15.10). We have shown that the eradication of reflux by an anti-reflux procedure is followed by the loss of "reflux induced spasm" with complete correction of symptoms.

# ESOPHAGEAL MOTOR DYSFUNCTION IN PATIENTS WITH GERD AND THE IMPLICATIONS FOR ANTIREFLUX SURGERY

The addition of ambulatory manometry to pH monitoring has provided further information on the pathophysiology of GERD that have implications for antireflux surgery. The clearance of gastroesophageal reflux is dependant on gravity, salivation, and on the motor function of the esophageal body.[26,27] Defective esophageal body function has been found in some patients with GERD and it is thought that this defect contributes to further exposure of the esophageal mucosa to gastric juice, resulting in a vicious cycle.[28,29] We have recently shown using ambulatory manometry that the motor function of the esophagus deteriorates with increasing degrees of mucosal damage. (Figure 15.11) This is particularly important when the esophagus is functioning, i.e., during meals. The normal response to eating is to increase in the amplitude of contractions and the prevalence of peristaltic wave forms.[30] However, some patients with GERD lose the ability to improve the esophageal motor function during meals. These patients are more susceptible to postoperative dysphagia which is one of the known complications that may follow an antireflux procedure.

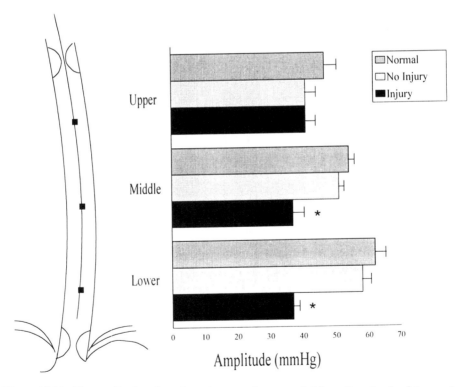

**Figure 15.11.** The amplitudes of esophageal contractions recorded from three levels of the esophagus during meals in normal subjects, patients with GERD and no mucosal injury, and those with GERD and mucosal injury. *$p<0.05$ vs normals and no-injury.

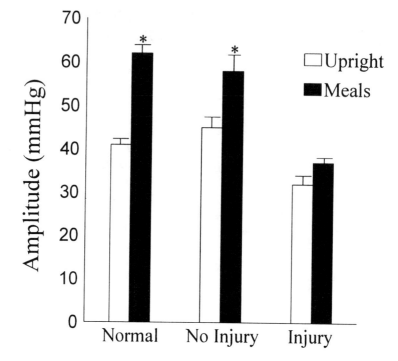

**Figure 15.12.** A graphic representation demonstrating the increase in the amplitude of contractions during meals for normal subjects and patients with no injury. Patients with mucosal injuring have lost the ability to do this. $p<0.05$ vs upright.

We have shown in a recent study of 45 patients with pH proven GERD, that 21 (46 percent) patients showed evidence of defective motility during meals. Only 7 of these were thought to have abnormal motor function on the basis of stationary manometry alone. Of interest is that this finding was most pronounced in patients with mucosal damage. (Figure 15.12) While the motor function of the esophagus may be expected to improve as the inflammatory changes subside after Nissen fundoplication, a more profound loss of contractility will not improve because muscle fibers have been destroyed.[29] Those patients who have a profound motor defect or who have lost the ability to improve motor function during eating should not have a complete fundoplication. Rather, they should have a partial fundoplication as this provides less of an outflow resistance to the passage of a food bolus.[31]

## POSTOPERATIVE ASSESSMENT

Antireflux surgery is different from the surgical extirpation of a diseased organ whose function is of no concern, since it will be destroyed with its removal. Rather, antireflux surgery is designed to improve the function of an organ that will remain in the patient and to provide complete and permanent relief of all symptoms and complications of gastro-esophageal reflux secondary to a mechanically defective lower esophageal sphincter. Ideally, the reconstructed cardia should permit the patient to swallow normally, belch to relieve gaseous distention, or vomit when necessary. The end result should restore the patient's esophageal function to normal, with no further need for medical, postural, or dietary therapy.

The problem with using symptoms as an indicator of success of antireflux surgery is that often a patient who has undergone surgery will not readily admit to the presence of postoperative symptoms. Gauging operative success only by the absence of the preoperative symptoms is inadequate. Similarly, the report of normal findings on a barium swallow is not a dependable criterion of success, since the symptoms of reflux or its complications may be present in a patient who has no radiologic evidence of a hiatus hernia or regurgitation of barium from the stomach into the esophagus. Rather, the success of a procedure depends on achieving the combination of (1) relief from symptoms, (2) objective evidence on 24 hour esophageal pH monitoring that esophageal acid exposure has been reduced to physiologic levels, and (3) evidence on esophageal manometry that the mechanical defect of the lower esophageal sphincter has been corrected. Consequently we ask all our patients who have antireflux surgery to have follow-up manometry and pH monitoring at 6, 18, and 36 months after the procedure.

The results of 100 consecutive primary Nissen's repairs performed over a 13 year period in patients with increased esophageal exposure to gastric juice documented by 24 hour esophageal pH monitoring showed that the actuarial success rate of the operation in controlling reflux symptoms over a 10 year period was 91 percent (Figure 15.13).[7] From the patients' perspective, 90 percent were satisfied with the results of the operation, and 92 percent would have the operation again if the decision had to be made over. Preoperatively the patients had a lower mean resting pressure of the lower esophageal sphincter, a shorter mean sphincter overall length, and a shorter mean sphincter abdominal length than healthy controls (Figure 15.14). In 36 patients studied postoperatively, the Nissen fundoplication reduced esophageal acid exposure to normal in all but six patients (Figure 15.15) four of whom were asymptomatic emphasizing the fact that symptoms alone are not a reliable guide to assess the success of an antireflux procedure. The operation also corrected the deficiencies of the lower esophageal sphincter (Figure 15.14). The ability of the Nissen fundoplication to restore the lower esophageal sphincter to normal can be appreciated by comparing the pre- and postoperative three-dimensional sphincter pressure profiles, a technique that integrates sphincter pressure, overall length, and abdominal length into one assessment (Figure 15.16).[32]

Failure of an antireflux procedure occurs when the patient, after the repair, is unable to swallow normally, experiences upper abdominal discomfort during and after meals, and has recurrence or persistence of heartburn, regurgitation, dysphagia, chest pain, nausea, or epigastric pain. The assessment of these symptoms and the selection of patients who need further surgical therapy remain challenging problems. Functional assessment of patients who have recurrent, persistent, or emergent new symptoms following a primary antireflux repair is critical to identifying the cause of failure. The results of these studies also identify those operative principles that are crucial to performing a successful primary antireflux procedure.

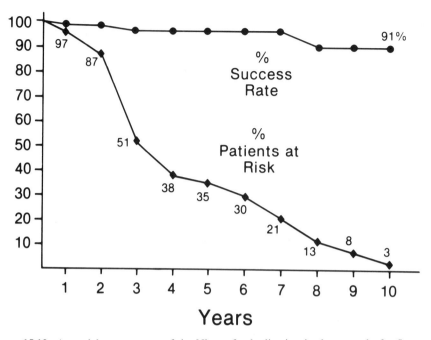

**Figure 15.13.** Actuarial success rate of the Nissen fundoplication in the control of reflux symptoms. The numbers on the lower curve represent the patients at risk for each subsequent yearly interval from which the actuarial curve was calculated. (From DeMeester TR, Bonavina L, Albertucci M: Nissen fundoplication for gastroesophageal reflux disease—evaluation of primary repair in 100 consecutive patients. *Ann Surg* 204:15, 1986, with permission.)

Recurrent heartburn and dysphagia are the most common symptoms for which a patient consults a surgeon following an antireflux repair. Dysphagia usually recurs or occurs immediately after the initial repair, whereas heartburn recurs more slowly. Postoperative esophageal manometry and 24 hour esophageal pH monitoring show that recurrent heartburn is associated with increased esophageal exposure to gastric juice resulting from a defective lower esophageal sphincter, a defective esophageal pump, or a combination of the two. Recurrent dysphagia is caused by incomplete relaxation of the lower esophageal sphincter, a defective esophageal pump, an anatomic stricture, stenosis, or a combination of these factors.[33]

A recent analysis of 63 patients requiring reoperation after a failed primary antireflux procedure showed that placement of the wrap around the stomach as opposed to the esophagus is the most frequent cause for failure of a primary antireflux procedure.[33] Other causes of failure are partial or complete breakdown of the wrap, herniation of the repair into the chest, and construction of a too tight or too long wrap. Attention to the technical details during construction of the primary procedure will avoid these failures in most instances. The critical role of preoperative esophageal function tests, i.e., 24 hour esophageal pH monitoring and esophageal manometry is underscored by the outcome of the five patients in this reoperated group, who had an antireflux procedure for a unrecognized underlying esophageal motor disorder.

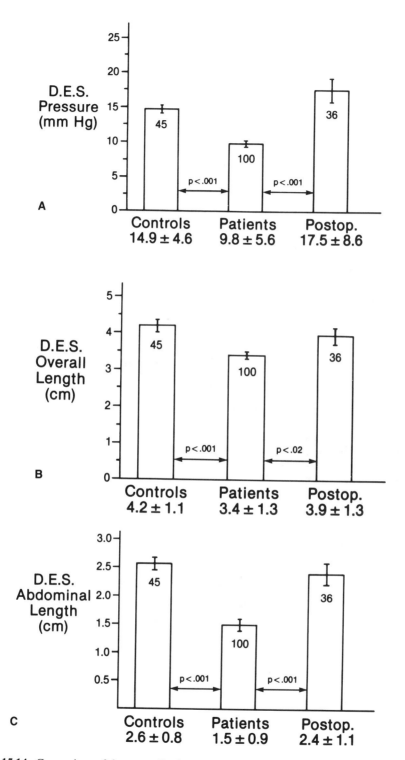

**Figure 15.14.** Comparison of the mean distal esophageal sphincter pressure (a), overall length (b), and abdominal length (c) measured in control subjects and in patients before and after fundoplication. D.E.S. = distal esophageal sphincter. (From DeMeester TR, Bonavina L, Albertucci M: Nissen fundoplication for gastroesophageal reflux disease—evaluation of primary repair in 100 consecutive patients. *Ann Surg* 204:19, 1986, with permission.)

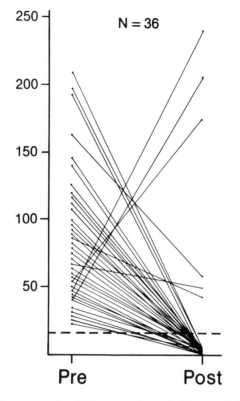

**Figure 15.15.** Pre- and postoperative 24 hour esophageal pH score in 36 patients. (From De-Meester TR, Bonavina L, Albertucci M: Nissen fundoplication for gastroesophageal reflux disease—evaluation of primary repair in 100 consecutive patients. *Ann Surg* 204:16, 1986, with permission.)

Abnormalities other than the function of the esophageal body and lower esophageal sphincter can also account for failure of a primary antireflux repair. Excessive duodenogastric reflux, cervical dysphagia, or gastric stasis may frequently cause symptoms following antireflux surgery. It appears that in patients with a diffuse foregut motor disorder, symptoms of gastroesophageal reflux, duodenogastric reflux, delayed gastric emptying, and/or an esophageal motor disorder may be present, with symptoms of one of these predominating. Correction of the predominant symptom results in the emergence of symptoms of the associated abnormality, which often requires additional surgery.

## THE ADVENT OF LAPAROSCOPIC SURGERY

The application of laparoscopic techniques to antireflux surgery has revealed that the procedure can be performed in the same manner as is done with an open incision. Thousands of patients have now been reported which show that the short-term results are excellent in

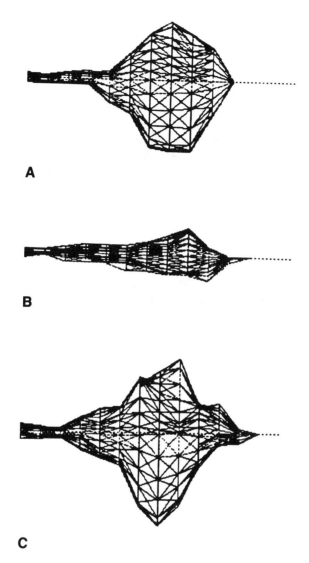

**Figure 15.16.** The three-dimensional lower esophageal sphincter pressure profile in a normal volunteer (a), a patient with a mechanically defective sphincter (b), and the same patient 1 year following Nissen fundoplication (c). The pressure profile was obtained by a stepwise pullback of eight radially oriented pressure transducers. Radial pressures along the gastroesophageal junction are plotted around an axis representing the gastric baseline pressure.

properly selected patients.[34-40] Since the techniques are the same we believe that long-term results of the laparoscopic procedure will parallel those of open antireflux surgery. Certainly the minimally invasive approach is attractive to both the patient and gastroenterologist, which may account for the trend toward the increased use of surgery in the treatment of this disease patients. It has further been shown to be cost-effective when com-

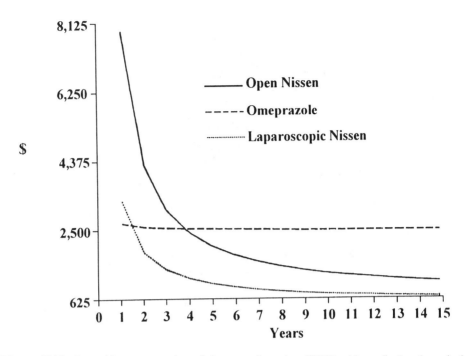

**Figure 15.17.** A graphic representation of the cost of treating GERD with medical and surgical therapy.

pared to chronic medical therapy with a point of crossover occurring at approximately 4 years for the open procedure and 2 years for the laparoscopic approach (Figure 15.17).

We must stress, as has been done for open surgery, that care be taken in selecting patients for surgery. The same indications for open surgery as discussed above, apply to the laparoscopic approach. There are a few further considerations specific to the laparoscopic approach. First, an assessment of esophageal length should be made preoperatively, since it is difficult to assess esophageal length intraoperatively when the diaphragms are displaced upward by the pneumoperitoneum. If the esophagus is short, the release of the pneumoperitoneum will place the repair under too much tension leading to either disruption of the repair or retraction of the repair into the chest. Herniation of the repair into the chest is now the most common complication following the laparoscopic approach with an international frequency of approximately 7 percent. Preoperative indications of a short esophagus include 1) a hiatal hernia that will not reduce while in the upright position, 2) the endoscopic measurement of greater than 4 cm from the diaphragmatic crura to the gastro-esophageal junction, 3) presence of a long columnar segment in Barrett's esophagus or an esophageal stricture. Second, if the patient has had a previously failed antireflux procedure an open approach should be encouraged. Scarring and anatomic deformities that result from the previous procedure can make reoperation difficult and perhaps dangerous via the laparoscopic approach.

# REFERENCES

1. DeMeester TR, Stein, HJ: Gastroesophageal reflux disease. In: Moody FG, Carey LC, Jones RC, et al., eds: *Surgical Treatment of Digestive Disease*, ed 2. Chicago: Year Book Medical Publishers, 1989:65.
2. Stein HJ, DeMeester TR: Who benefits from antireflux surgery? *World J Surg* 1991 (in press).
3. DeMeester TR, Wang CI, Wernly JA, et al: Technique, indications and clinical use of 24-hour esophageal pH monitoring. *J Thorac Cardiovasc Surg* 79:656–667, 1980.
4. DeMeester TR, Stein HJ, Fuchs KH: Diagnostic studies in the evaluation of the esophagus— physiologic diagnostic tests. In: Zuidema GD, Orringer MB, eds: *Surgery of the Alimentary Tract, ed 3*. Philadelphia: W. B. Saunders Company, 1991:94.
5. DeMeester TR: Prolonged oesophageal pH monitoring. In: Read NW, ed: *Gastrointestinal Motility: Which Test?* Petersfield, England: Wrightson Biomedical Publishing Ltd., 1989:41.
6. Zaninotto G, DeMeester TR, Schwizer W, et al: The lower esophageal sphincter in health and disease. *Am J Surg* 155:104–111, 1988.
7. DeMeester TR, Bonavina L, Albertucci, M: Nissen fundoplication for gastroesophageal reflux disease—evaluation of primary repair in 100 consecutive patients. *Ann Surg* 204:9–20, 1986.
8. Stein HJ, Barlow AP, DeMeester TR, et al: Complication of gastroesophageal reflux disease. Role of the lower esophageal sphincter, esophageal acid and acid/alkaline exposure, and duodenogastric reflux. *Ann Surg* 216(1):35–43, 1992.
9. Mason RJ, Lund RJ, Peters JH, et al: A Nissen prevents shortening of the sphincter during gastric distension. *Arch Surg* (In Press):
10. Costantini M, Zaninotto G, Anselmino M, Boccu C, Nicoletti L, Ancona E: The role of a defective lower esophageal sphincter in the clinical outcome of treatment for gastroesophageal reflux disease. *Arch Surg* 131:655–659, 1996.
11. Stein HJ, Eypasch EP, DeMeester TR, et al: Circadian esophageal motor function in patients with gastroesophageal reflux disease. *Surgery* 108:769–778, 1990.
12. DeMeester TR, Johansson K-E, Franze I, et al: Indications, surgical technique, and long-term functional results of colon interposition or bypass. *Ann Surg* 208:460, 1988.
13. DeMeester TR, Attwood SEA, Smyrk TC, et al: Surgical therapy in Barrett's esophagus. *Ann Surg* 212:528–542, 1990.
14. DeMeester TR, Bonavina L, Iascone C, et al: Chronic respiratory symptoms and occult gastroesophageal reflux. *Ann Surg* 211:337, 1990.
15. DeMeester TR, O'Sullivan GC, Bermudez G, et al: Esophageal function in patients with angina-type chest pain and normal coronary angiograms. *Ann Surg* 196:488–498, 1982.
16. Fuchs KH, Hinder RA, DeMeester TR, et al: Computerized identification of pathologic duodenogastric reflux using 24-hour gastric pH monitoring. *Ann Surg* 1991 213; 13–20.
17. Stein HJ, Hinder RA, DeMeester TR, et al: Clinical use of 24-hour gastric pH monitoring vs. O-diisopropyl iminodiacetic acid (DISIDA) scanning in the diagnosis of pathologic duodenogastric reflux. *Arch Surg* 125:966–971, 1990.
18. DeMeester TR, Fuchs KH, Ball CS, et al: Experimental and clinical results with proximal end-to-end duodenojejunostomy for pathologic duodenogastric reflux. *Ann Surg* 206:414–426, 1987.
19. Paterson WG, Murat BW: Combined ambulatory esophageal manometry and dual-probe pH-metry in evaluation of patients with chronic unexplained cough. *Dig Dis Sci* 39:1117–1125, 1994.
20. DeMeester TR, Bonavina L, Iascone C, Courtney JV, Skinner DB: Chronic respiratory symptoms and occult gastroesophageal reflux. *Ann Surg* 211:337–345, 1990.
21. Patti MG, Debas HT, Pellegrini CA: Esophageal manometry and 24-hour pH monitoring in the diagnosis of pulmonary aspiration secondary to gastroesophageal reflux. *Am J Surg* 163:401–406, 1992.

22. Pellegrini CA, DeMeester TR, Johnson LF, et al: Gastroesophageal reflux and pulmonary aspiration: Incidence, functional abnormality, and results of surgical therapy. *Surgery* 86:110–119, 1979.

23. Gomez R, Moreno E, Seoane J, Volwald P, Calle A, Moreno C. Esophageal pH monitoring of postprandial gastroesophageal reflux. Comparison between healthy subjects, patients with gastroesophageal reflux and patients treated with Nissen fundoplication. *Dig Dis* 11:354–362, 1993.

24. Kruse Anderson S, Wallin L, Madsen T: Acid gastro-oesophageal reflux and oesophageal pressure activity during postprandial and nocturnal periods. A study in subjects with and without pathologic acid gastro-oesophageal reflux. *Scand J Gastroenterol* 22:926–930, 1987.

25. Katz LC, Just R, Castell DO. Body position affects recumbent postprandial reflux. *J Clin Gastroenterol* 18:280–283, 1994.

26. Bremner RM, Hoeft SF, Costantini, et al: Pharyngeal swallowing: The major factor in clearance of esophageal reflux episodes. *Ann Surg* 218(3): 364–370, 1993.

27. Corazziari E, Bontempo I, Anzini F, Torsoli A: Motor activity of the distal oesophagus and gastrooesophageal reflux. *Gut* 25:7–13, 1984.

28. Joelsson BE, DeMeester TR, Skinner DB, et al: The role of the esophageal body in the antireflux mechanism. *Surgery* 92:417–424, 1982.

29. Stein HJ, Bremner RM, Jamieson J, DeMeester TR. Effect of Nissen fundoplication on esophageal motor function. *Arch Surg* 127:788–791, 1992.

30. Bremner RM, Costantini M, DeMeester TR, et al: Normal esophageal function: a study using ambulatory esophageal manometry. *Am J Gastroenterol* 1996; In Press:

31. Kauer WK, Peters JH, DeMeester TR, Heimbucher J, Ireland AP, Bremner CG: A tailored approach to antireflux surgery. *J Thorac Cardiovasc Surg* 110:141–146, discussion 146–147, 1995.

32. Stein HJ, DeMeester TR, Perry RP: Three-dimensional lower esophageal sphincter pressure profile in gastroesophageal reflux disease. *Ann Surg* 214:374–384, 1991.

33. Collard JM, DeMeester TR: Correction of failed antireflux procedures based on preoperative functional assessment and intraoperative findings (in preparation).

34. Peters JH, DeMeester TR: Indications, principles of procedure selection, and technique of laparoscopic Nissen fundoplication. *Semin Lap Surg* 2:27–44, 1995.

35. Hinder RA, Raiser F, Katada N, McBride PJ, Perdikis G, Lund RJ: Results of Nissen fundoplication. A cost analysis. *Surg Endosc* 9:1328–1332, 1995.

36. Anvari M, Allen C, Borm A: Laparoscopic Nissen fundoplication is a satisfactory alternative to long-term omeprazole therapy. *Br J Surg* 82:938–942, 1995.

37. Watson DI, Jamieson GG, Mitchell PC, Devitt PG, Britten Jones R. Stenosis of the esophageal hiatus following laparoscopic fundoplication. *Arch Surg* 130:1014–1016, 1995.

38. Hunter JG, Swanstrom L, Waring JP. Dysphagia after laparoscopic antireflux surgery. The impact of operative technique. *Ann Surg,* 224:51–57, 1996.

39. Gotley DC, Smithers BM, Rhodes M, Menzies B, Branicki FJ, Nathanson L. Laparoscopic Nissen fundoplication—200 consecutive cases. *Gut* 38:487–491, 1996.

40. Peters JH, Heimbucher J, Kauer WK, Incarbone R, Bremner CG, DeMeester TR. Clinical and physiologic comparison of laparoscopic and open Nissen fundoplication [see comments]. *J Am Coll Surg* 180:385–393, 1995.

# 16

# Pediatric Applications of pH Monitoring

Carlo DiLorenzo, M.D.
Paul E. Hyman, M.D.

Daily regurgitation is within the range of expected behaviors in healthy infants. Between 2 and 10 months of age approximately 40 percent of normal infants regurgitate at least once every day.[1] Evaluation of these infants reveals no anatomic, metabolic, infectious, or neurologic abnormalities and regurgitation resolved spontaneously in most cases by 18 months of age.[2] In contrast, adults rarely experience spontaneous prolonged remission of Gastroesophageal reflux (GER) symptoms. Factors contributing to the age-related improvement include (but may not be limited to) gradual increase in lower esophageal sphincter pressure during the first months of life,[3] passage to an upright position, and transition to a diet containing solid foods.[4]

In most cases, benign and transient GER in infancy and childhood is occasionally associated with significant morbidity and even mortality.[5] Consequences of GER in pediatric patients are detailed in Table 16.1. Termed gastroesophageal reflux—associated diseases (GERD), they include recurrent vomiting with caloric losses and subsequent failure to thrive, esophagitis, unusual posturing (Sandifer's syndrome), pain, anemia, stricture formation, aspiration pneumonia, apnea, and reactive airway disease. In infants, behaviors temporally associated with GERD identified by simultaneous recording of pH and split-screen video included crying, yawning, burping, hiccup, stridor, cough, and sneezing.[6] Thus, signs and symptoms associated with gastroesophageal reflux in infants and children may differ from those reported in adults.

Because diversity of clinical presentations in infants and children often makes the diagnosis of GERD difficult, emphasis has been given to the identification of the best test for the diagnosis of GERD. In children who vomit, barium swallow is essential to evaluate anatomy, but it is not useful for assessing GERD. Barium swallow has been shown to result in false positive diagnoses in up to 31 percent and false negative diagnoses in up to 14 percent of children.[7] Accurate diagnosis depends on the experience of the radiologist, the amount of barium used, and the time spent performing the study.[8] The water siphon test is a refinement of the barium esophagogram used in infants,[9] but it is felt by many to have an unacceptably high false-positive rate.

TABLE 16.1. GASTROESOPHAGEAL REFLUX-ASSOCIATED DISEASES
            IN PEDIATRICS

|  | Symptoms | Frequency | Sequelae |
|---|---|---|---|
| 1. Recurrent vomiting | Emesis | Common | Failure to thrive Disordered mother-child interactions |
| 2. Pulmonary symptoms | Wheezing | Common | Contributes to asthma |
|  | Aspiration | Common in developmental delay | Pneumonia |
|  | Apnea | Uncommon | Sudden infant death syndrome |
| 3. Esophagitis | Pain, irritability | Common | Feeding problems anorexia |
|  | Hematemesis | Uncommon | Anemia |
|  | Posturing (Sandifer's syndrome) | Uncommon | Barretts esophagus Stricture |

Another test that is used in pediatric more than in adult patients to detect GER is the gastroesophageal scintiscan (milk scan).[10-11] Its advantages over pH monitoring are that it requires no intubation, detects nonacid as well as acid reflux, and identifies pulmonary aspiration. Pitfalls are the short duration of the scintiscan and the detection of only post-prandial reflux episodes, which are felt by many to be a normal physiologic event.[12] A study comparing the accuracy of different methods to diagnose reflux in children concluded that pH monitoring is the test that correlates best with GER.[13]

A survey by the Patient Care Committee of the North American Society of Pediatric Gastroenterology and Nutrition (NASPGN) reported that 87 percent of the members used pH monitoring to diagnose GER. Respondents performed an average of 80 intraesophageal pH studies per year. There was consensus that intraesophageal pH monitoring for 18 to 24 hours is the method of choice for evaluating children in whom GER is suspected.

# APPROACH TO pH MONITORING IN INFANTS AND YOUNG CHILDREN

The ideal duration for pH monitoring has been a matter of debate. If one wishes only to know whether a patient has excessive amounts of reflux, there may be no need for a 24-hour study. It has been shown that a test session as short as 3 hours is as sensitive and specific as all day monitoring to diagnose GER.[14] Short-term monitoring is performed with the patient in the recumbent position and in the fasting state to avoid the frequent episodes of reflux that occur normally in the immediate postprandial period.[15] Limiting the pH recording to postprandial periods reduces the ability of the test to quantitate reflux.[16] The administration of a standard acidified meal (300 ml per 1.73 sq m of 0.1 N hydrochloric acid) is thought to increase the sensitivity of the test and further shortens its duration to 30 or 60 minutes (the Tuttle test).[17] However, gastric distention leads to inappropriate lower esophageal sphincter relaxation,[18] predisposing the subject to episodes of reflux and casting doubt on the value of a test in which a large volume is infused involuntarily into the stomach.

Prolonged pH monitoring is used to correlate reflux episodes with discrete symptoms or abnormal behaviors. It also identifies the temporal pattern of GER and its relationship to type of feeding, body position, and physical activity, which helps to plan a therapeutic approach.

The test should be performed when the patient is in his or her usual state of health and not during an acute illness. In older children and adults, pH monitoring is often performed as an outpatient study. When this is not possible, or when the child is less than 10 years old, try to wait for the child to be at his or her best before performing the test in the hospital. Results obtained from a combative, intubated, stressed, or sedated child cannot be assumed to be characteristic of results from the same child studied in a more normal state. The aim of the study in most cases is to anticipate if reflux will persist after the child leaves the hospital and resumes his or her previous lifestyle.

After insertion, the pH probe is well tolerated by almost all infants less than 8 to 12 months of age. As infants age beyond this point, however, higher percentages do not accept it. Occasionally children may need to wear mittens or have soft restraints placed on the extremities to prevent them from pulling out the probe.

One contraindication to pH monitoring is the presence of severe respiratory distress, because the placement of the electrode in the nares can cause further respiratory compromise in small infants, who are obligate nose breathers.

When performing pH monitoring in newborns and infants, problems are encountered that are peculiar to the pediatric age group. The diameter of the electrode used for adults is too large for a newborn. Small glass microelectrodes and antimony electrodes have partially overcome this problem (Figure 16.1). They have a 1.3 mm diameter, which allows placement even in a premature baby. They are as accurate as larger electrodes, although they are fragile and have a shorter life. Unlike some adult size glass electrodes that have an internal reference electrode, these miniaturized electrodes need an external reference electrode. The reference electrode attached to the skin may be dislodged by a combative or inquisitive child. Careful taping to secure both recording and reference electrodes and supervision by an adult become particularly important in toddlers and young children. Recently, antimony electrodes with a diameter of approximately 2.0 mm and an internal reference electrode have been marketed (Synectics, Stockholm, Sweden, M.I.C. Solothurn, Switzerland). pH data were not influenced by the recording devices, the age of the electrodes, or the skin potential differences induced by the person calibrating the probes.[19]

Location of the catheter tip is an arbitrary but critical factor in evaluating results of a pH study: the closer the probe is to the cardia, the more reflux episodes are recorded. In adults, the electrode is usually placed 5 cm above the lower esophageal sphincter. In pediatric patients the electrode is usually placed 87 percent of the distance from the nares to the lower esophageal sphincter (the rationale being that 5 cm is 13 percent of the standard adult esophageal length from the teeth to the lower esophageal sphincter). Esophageal length from the nares to the middle of the lower esophageal sphincter can be estimated

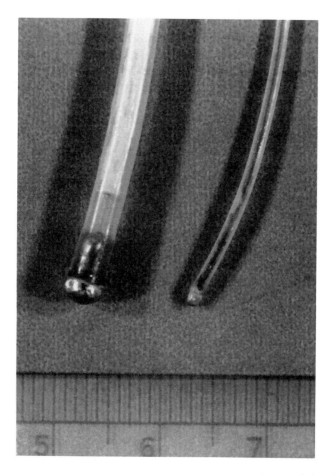

**Figure 16.1.** Electrode used for adult (on the left) and for pediatric (on the right) pH monitoring studies. Both required an external reference electrode.

from the patient's height using the Strobel formula: $5 + (0.252 \times \text{height in centimeters})$.[20] This formula is dependable in children who do not have hiatus hernia. The most accurate placement of the electrode follows measurement of nares-lower esophageal sphincter distance by esophageal manometry, but this equipment is not widely used by most pediatricians. The European Society for Pediatric Gastroenterology and Nutrition has recommended fluoroscopy to position the pH electrode. The tip of the electrode should overlie the third vertebral body above the diaphragm throughout the respiratory cycle.[21]

Another problem frequently encountered when performing pH monitoring in infants is the buffering effect of breast milk or formula on gastric acid. When milk is given, the intragastric pH rises above 4.0, and reflux episodes may be masked. To avoid this source of error, some investigators use apple juice (which has pH around 4) or another clear liquid during the test session. Apple juice and other clear liquids empty from the stomach more rapidly than complex liquids or solids, and so do not simulate a real meal. Since the frequency, acidity, composition, and volume of the feedings influence the incidence of reflux,[21] each testing center should evaluate the results of its studies based on its own protocol. Body position also affects reflux in children more than in adults. For example, reflux is less common in the prone than in the supine position or in the infant seat.[22] There is also an increase in occurrence of physiologic reflux episodes during awake periods or periods of physical activity. As a consequence, published normal values[16,23–25] should be used with caution if study conditions differ. pH monitoring data from children without GERD are reported in Table 16.2. Differences exist among data reported by the different authors, probably reflecting the study population variation in age, feeding patterns, and positioning. In some cases[16,25] the "normal" values were from children hospitalized for GER evaluations who subsequently were found to have a different cause for their symptoms. Very hot and very cold beverages and food are avoided during the testing because the electrodes are temperature sensitive.[21]

Considerable differences of opinion exists concerning the interpretation of the results. The evaluation parameters most used in pediatrics to evaluate the data are based on the work of Johnson and DeMeester.[26] They are the percentage of the investigation time with pH less than 4.0, the duration of the longest episode with pH less than 4.0, the number of episodes with pH less than 4.0, and the number of episodes longer than 5 minutes with pH less than 4.0. Total esophageal exposure time and the number of episodes requiring more than 5 minute to clear have been reported to be the parameters that best differentiate children with different severity of reflux and best predict the presence of esophagitis.[16] A parameter that takes into account both the acidity and duration of GER is the "area under pH 4."[27] It originates from the assumption that peptic damage to the esophagus is more severe with a more acidic refluxate, and therefore by using this parameter one avoids the problem of weighing all the refluxes with pH less than 4.0 in the same way. It has also been noted that about 50 percent of reflux episodes occurring in infants during sleep are characterized by gradual drift of pH just above or below 4.0.[15] Therefore, measuring the percentage of time the pH varied around pH 4.0 (e.g., between 4.25 and 3.75, the "oscillatory index") has been proposed.[28]

In pediatrics, the reproducibility of the test appears to be excellent. When performed on two consecutive days in similar circumstances, the two tests had a positive coefficient of correlation for every variable evaluated, ranging from 0.88 to 0.98.[29]

### TABLE 16.2. VALUES FOR 24 HOUR ESOPHAGEAL pH MONITORING IN INFANTS AND CHILDREN WITHOUT GER*

| | Vandenplas[24] (N = 92; age < 15d) | Boix-Ochoa[25] (N = 20; X̄ age = 19 mo) | Sondeheimer[26] (N = 11; X̄ age = 61 mo) | Cucchiara[16] (N = 63; X̄ age = 24 mo) |
|---|---|---|---|---|
| Reflux index | 1.2 ± 0.9 | 1.9 ± 1.6 | 3.2 ± 1.9 | 1.4 ± 1 |
| Number of reflux episodes/24h | 7.7 ± 6.5 | 10.6 ± 8.8 | 19.2 ± 9.6 | 11 ± 8.5 |
| Number reflux episodes > min/24h | 0.6 ± 0.5 | 1.7 ± 2.0 | 3.4 ± 2.0 | 0.6 ± 0.9 |
| Duration of longest reflux episode (min) | 3.8 ± 1.9 | 8.1 ± 7.2 | — | 6.3 ± 5.9 |

*Values represent mean ±1 standard deviation

Prolonged monitoring of gastric pH constitutes another application of pH monitoring. Especially in seriously ill patients in intensive care units, it is often important to achieve a continuous suppression of acid secretion. The pH probe provides accurate and constant information about the gastric pH. Monitoring of gastric pH has also proved useful in assessing efficacy of antacid medications in infants[30] and children with peptic ulcer disease.

# APPLICATIONS OF pH MONITORING IN CHILDREN

Because GERD is common in children and pH monitoring appears to be a safe, sensitive, and reproducible method to detect reflux, there are many clinical conditions in which this test could be used in children. The test is unnecessary when the diagnosis is clear from the clinical presentation. For example, in children who fail to thrive because of recurrent vomiting, documentation of reflux with a pH study is unnecessary. The main indications for pH monitoring in the pediatric population are the documentation of reflux in children with atypical symptoms or abnormal behaviors. In a recently published position statement, a North American Society for Pediatric Gastroenterology and Nutrition-sponsored consensus committee considered laryngeal symptoms such as chronic cough, sore throat, and hoarseness, recurrent pneumonia, unexplained recurrent apnea and bradycardia in infancy, and atypical chest pain as clinical situations when pH monitoring is generally useful.[32] In other situations, such as infant irritability, intractable crying, food refusal and asthma, pH monitoring was felt to be helpful in selected cases. It was felt that pH monitoring had little role in the evaluation of childhood dysphagia.[32]

## Case No. 1

T.K. was born at gestational age 27 weeks, weighing 1.28 kg. The mother had a past history of endometriosis and genital herpes. A cesarean section was performed secondary to breech presentation and herpes in the fetus. The neonate required mechanical ventilation for several days. As soon as he was weaned to room air, he began to suffer from apnea and bradycardia several times a day. These events were treated successfully with aminophylline and phenobarbital. He was discharged at 2 months of age, weighing 1.96 kg. Three days after discharge he began to have apneic episodes accompanied by bradycardia in response to physical stimulation. These episodes occurred during and after feedings. The patient was readmitted to the hospital. Work-up for sepsis was negative. His phenobarbital level was adequate, and blood chemistry results were normal. An electroencephalogram showed seizure-like activity, although a consultant in neurology felt that the seizure activity was not the cause of the apnea and bradycardia. Ultrasonography of the head showed mild lateral ventricular dilation. The patient was anemic but continued to have apneic episodes after transfusion. A pH probe study revealed five episodes of reflux during a 24-hour test session. However, two of these episodes were prolonged, with the longest lasting 44 minutes. Five seconds after the beginning of one of these reflux episodes, the patient experienced a 12 second episode of central apnea accompanied by bradycardia and a decrease in transcutaneous oxygen saturation (Figure 16.2). This episode resolved spontaneously, and oxygen saturation returned to normal. This episode was the only apnea and bradycardia occurring during the test session. The patient was treated for the GER with postprandial positioning prone, head elevated, and a prokinetic drug. The number of apneic and bradycardic episodes decreased markedly. The infant

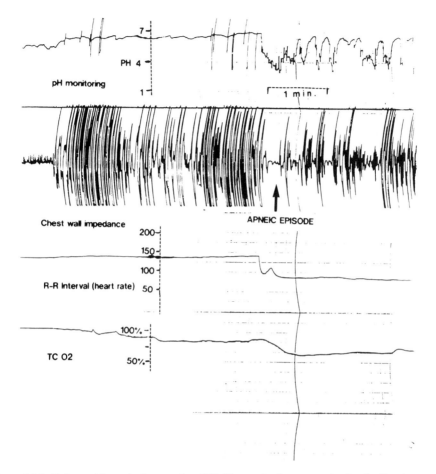

**Figure 16.2.** Polygraphic study from patient T.K. The study shows an episode of reflux (top panel) followed by apnea (second panel), bradycardia (third panel), and a drop in oxygen saturation (bottom panel).

was discharged from the hospital 2 weeks later with home apnea monitoring equipment. He had no further episodes.

## Case No. 2

J.V. was a 13 month old formerly premature infant with bronchopulmonary dysplasia, reactive airway disease, and gastroesophageal reflux documented by upper gastrointestinal radiography and scintiscan 2 months prior to hospital admission. The patient had a history of daily emesis and several episodes of aspiration pneumonia. In spite of outpatient treatment with metoclopramide and dietary changes including thickened feedings, he continued to vomit. He was admitted to the hospital for fever, cough, and increasing emesis. He had had multiple prior admissions at other institutions for pneumonia. He was also extremely underweight, with an admission weight of 6.11 kg (under the 5th percentile).

On physical examination, the patient was not in distress. Head and neck examination was normal. Lung examination revealed expiratory rales, greater on the left side, without wheezing or retractions. Abdominal examination revealed the liver palpable 3 cm below the right costal margin and no palpable spleen. Neurologic examination revealed developmental delay. A 24-hour pH study was abnormal. There were 29 reflux episodes. The longest lasted 54 minutes. The majority of episodes occurred within 2 hours of feeding, but most of the long episodes occurred during sleep.

Because of the delayed growth and recurrent respiratory symptoms despite attempts at medical management, the patient underwent a Nissen fundoplication and gastrostomy tube placement 2 weeks after hospital admission. He had a stormy postoperative course, which included obstruction of the jejunum by adhesions, resulting in partial small bowel infarction, perforation, and intra-abdominal abscess. He responded to supportive therapy and antibiotics and was ultimately discharged 2 months later.

After discharge, his respiratory symptoms improved. He did not have further aspiration pneumonias. His rate of growth improved, and over the next 3 months he gained an average of 35 g/day (normal 20 to 30 g/day).

The relationship between GER and events such as apneic spells, sleep disorders, and periodic breathing is a matter of great debate in the pediatric literature.[31,32] The first reports of an association between GER and recurrent apnea appeared in 1977.[33] Subsequent studies have generally confirmed that GER is a frequent event in infants who present with apneic spells, but there is as yet no convincing proof that GER is the cause of more than a small fraction of infant apneic episodes. There are no controlled studies in patients with apnea and GER comparing outcome of infants who were treated for reflux to outcome of those who were not. It has been suggested that prolonged recording of pH in the proximal esophagus is better able to identify infants whose reflux causes apnea.[34] Until there are more careful studies of the relationship between esophageal acidification and respiratory control, caution should be exercised before attributing the cause of recurrent apnea to GER unless there is a close temporal relationship between a reflux episode and the respiratory event. When a causal relationship between gastroesophageal reflux and apnea is suspected, the pH monitoring test should be part of a multi-channel pneumocardiography test.[32]

## Case No. 3

J.R. was a 14 year old boy admitted for intractible pain, anorexia, and weight loss. He was found to have a duodenal ulcer. Therapy with intravenous cimetidine was associated with confusion, headache, and dysphoria. Treatment with antiacids, sucralfate, bismuth, propantheline, and isopropamide was ineffective. Diagnostic studies showed basal acid output was 20 mmol/hr and maximal acid output was 40 mmol/hr; fasting gastrin ranged between 40 and 70 pg/ml; secretin test and $Ca^2$ infusion followed by the secretin test were normal. Infusion of atropine reduced basal acid output by 80 percent. A computed tomographic scan and magnetic resonance imaging of the abdomen were negative. The patient was started on omeprazole because of his gastric hypersecretory condition. The dose of omeprazole was increased by 20 mg increments every 5 days up to 120 mg/day, using 24-hour gastric pH monitoring as a guide to reach the minimal effective dose. The ulcer healed with this dose, and the patient became asymptomatic. After 6 months of therapy, omeprazole was reduced to 80 mg/day. Within several weeks the patient again complained of abdominal pain. Endoscopy showed recurrence of the ulcer, and pH monitoring demonstrated that gastric acid pH was below 2.0 more than 80 percent of the time (Figure 16.3). Again, pH monitoring showed that a dose of 120 mg/day was needed to control gastric acid secretion.

This case illustrates another use of pH studies in children: monitoring the effects of drug therapy. In pediatrics there are no drugs approved by the Federal Drug Administration for therapy of peptic ulcer disease. In many cases the drugs must be titrated to reach the minimum effective dose in each subject. The use of a pH probe is helpful in monitoring therapy and allows repeated studies with minimal patient discomfort and high acceptability.

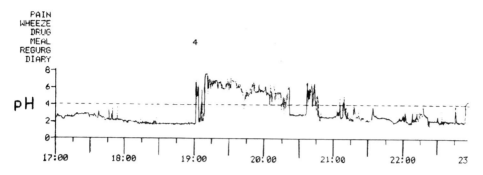

**Figure 16.3.** Part of the intragastric pH monitoring from patient J.R. The study was performed while the patient was taking omeprazole, 80 mg/day. The recording showed that the gastric pH was above 4.0 only in the hour following the meal (indicated by the number 4).

## SUMMARY

Symptoms of GER commonly cause clinical problems in infants and children. To a great extent, infant regurgitation is a physiological event that improves with maturation. Esophageal pH monitoring can be most helpful in documenting GER in children with atypical symptoms or abnormal behavior, and in clarifying the reasons for treatment failures if GER persists despite medical or surgical therapy.

## REFERENCES

1. Orenstein SR, Shalaby TS, Cohn JF. Gastroesophageal reflux symptoms in 100 normals: diagnostic validity of the infant gastroesophageal reflux questionnaire. *J Pediatr Gastroenterol Nutr* 21:33, 1995.
2. Carre IJ: The natural history of partial thoracic stomach ("hiatal hernia") in children. *Arch Dis Child* 34:344–352, 1953.
3. Boix-Ochoa J, Canals J: Maturation of the lower esophagus. *J Pediatr Surg* 11:749–756, 1976.
4. Orenstein SR, Magill HL, Brooks P: Thickening of infant feedings for therapy of gastroesophageal reflux. *J Pediatr* 110:181–186, 1987.
5. Herbst JJ: Gastroesophageal reflux. *J Pediatr* 98:859–870, 1981.
6. Feranchak AP, Orenstein SR, Cohn JF. Behaviors associated with onset of gastroesophageal reflux episodes in infants: prospective study using split screen video and pH probe. *Clin Pediatr* 33:654–662, 1994.
7. Meyers WF, Roberts CC, Johnson DG, et al: Value of tests for evaluation of gastroesophageal reflux in children. *J Pediatr Surg* 20:515–520, 1985.
8. Jewett TC Jr, Siegel M: Hiatal hernia and gastroesophageal reflux. *J Pediatr Gastroenterol Nutr* 3:340–345, 1984.
9. Blumhagen JD, Christie DL: Gastroesophageal reflux in children: Evaluation of the water siphon test. *Radiology* 131:345–340, 1979.
10. Heyman S, Kirkpatrick JA, Winter HS, et al: An improved radionuclide method for diagnosis of gastroesophageal reflux and aspiration in children (milk scan). *Radiology* 131:479–482, 1979.

11. Jona JZ, Sty JR, Glicklich M: Simplified radioisotope technique for assessing gastroesophageal reflux in children. *J Pediatr Surg* 16:114–117, 1981.

12. Jolley SG, Herbst JJ, Johnson DG, et al: Postcibal gastroesophageal reflux in children. *J Pediatr Surg* 16:487–490, 1981.

13. Arasu TS, Willie R, Fritzgerald JR: Gastroesophageal reflux in infants and children-comparative accuracy of diagnostic methods. *J Pediatr* 96:798–803, 1980.

14. Reyes HM, Ostrovsky E, Radhakrishnan J: Diagnostic accurancy of a 3 hour continuous intraluminal pH monitoring of the lower esophagus in the evaluation of gastro-esophageal reflux in infancy. *J Pediatr Surg* 17:626–631, 1982.

15. Sondheimer JM: Esophageal pH monitoring. In: Walker WA, Durie PR, Hamilton JR, et al, eds: *Pediatric Gastrointestinal Disease*. Philadelphia: BC Decker Inc, 1990:1331.

16. Cucchiara S, Staiano AM, Gobio Casali L, et al: Value of the 24 hour intraesophageal pH monitoring in children. *Gut* 31:129–133, 1990.

17. Euler AR, Ament ME: Detection of gastroesophageal reflux in the pediatric age patient by esophageal intraluminal pH probe measurement (Tuttle test). *Pediatrics* 60:65–68, 1977.

18. Holloway RH, Hongo M, Berger K, et al: Gastric distention: A mechanism for postprandial gastroesophageal reflux. *Gastroenterology* 89:779–784, 1985.

19. Vandenplas Y, Goyvaerts H, Helven R: Do esophageal pH monitoring data depend on recording equipment and probes? *J Pediatr Gastroenterol Nutr* 10:322–326, 1990.

20. Strobel CT, Byrne WJ, Ament ME, et al: Correlation of esophageal lengths in children with height: Application of the Tuttle test without prior esophageal manometry. *J Pediatr* 94:81–86, 1979.

21. Vandenplas Y, Belli D, Boige N, et al. A standardized protocol for the methodology of esophageal pH monitoring and interpretation of the data for the diagnosis of gastroesophageal reflux. *J Pediatr Gastroenterol Nutr* 14:467–71, 1992.

22. Vandenplas Y, Loeb H: The interpretation of esophageal pH monitoring data. *Eur J Pediatr* 149:598–602, 1990.

23. Orenstein SR, Whitington PF: Positioning for prevention of infant gastroesophageal reflux. *J Pediatr* 103:534–537, 1983.

24. Vandenplas Y, Sacré-Smits L: Continuous 24-hour esophageal pH monitoring in 285 asymptomatic infants 0–15 months old. *J Pediatr Gastroenterol Nutr* 6:220–224, 1987.

25. Boix-Ochoa J, Lafuente JM, Jil-Vernet JM: 24-hour esophageal pH monitoring in gastroesophageal reflux. *J Pediatr Surg* 15:74–78, 1980.

26. Sondheimer JM, Haase GA: Simultaneous pH recording from multiple sites in children with and without distal gastroesophageal reflux. *J Pediatr Gastroenterol Nutr* 7:46–51, 1988.

27. Johnson LF, Deemester TR: Twenty-four hour pH monitoring of the distal esophagus, a quantitative measure of gastroesophageal reflux. *Am J Gastroenterol* 62:325–332, 1974.

28. Vandenplas Y, Franckx-Goossens A, Pipeleers-Marichal M, et al: Area under pH 4: Advantages of a new parameter in the interpretation of esophageal pH monitoring data in infants. *J Pediatr Gastroenterol Nutr* 9:34–39, 1989.

29. Vandenplas Y, Lepoudre R, Helven R: Dependability of esophageal pH-monitoring data in infants on cutoff limits: The oscillatory index. *J Pediatr Gastroenterol Nutr* 11:304–309, 1990.

30. Vandenplas Y, Helven R, Goyvaerts H, et al: Reproducibility of continuous 24 hour oesophageal pH monitoring in infants and children. *Gut* 31:374–377, 1990.

31. Stuphen JL, Dillard VL, Pipan ME: Antacids and formula effects on gastric acidity in infants with gastroesophageal reflux. *Pediatrics* 78:55–57, 1986.

32. Colletti RB, Christie DL, Orenstein SR. Indications for pediatric esophageal pH monitoring: Statement of the North American Society for Pediatric Gastroenterology and Nutrition (NASPGN). *J Pediatr Gastroenterol Nutr* 21:253–262, 1995.

33. Walsh JK, Farrell ML, Keenan WJ: Gastroesophageal reflux in infants: Relation to apnea. *J Pediatr* 99:197–201, 1981.

34. Payton JY, Macfadyen U, Williams A, et al: Gastroesophageal reflux and apnoeic pauses during sleep in infancy-no direct relation. *Eur J Pediatr* 149:680–686, 1990.

35. Leape LL, Holder TM, Franklin JD, et al: Respiratory arrest in infants secondary to gastro-esophageal reflux. *Pediatrics* 60:924–927, 1977.

36. Haase GM, Ross MN, Gance-Cleveland B, et al: Extended four-channel esophageal pH monitoring: The importance of acid reflux pattern in the middle and proximal levels. *J Pediatr Surg* 23:32–37, 1988.

# 17

# Duodenogastroesophageal Reflux Alias "Alkaline" or "Bile" Reflux

## Michael F. Vaezi, Ph.D., M.D.

The term duodenogastroesophageal reflux (DGER) refers to regurgitation of duodenal content through the pylorus into the stomach, with subsequent reflux into the esophagus. Previously, the terms "bile reflux" and "alkaline reflux" were used to describe this process. However, duodenal contents contain more than just bile, and recent studies show that the term "alkaline reflux" is a misnomer since pH $> 7$ does not correlate with reflux of duodenal contents.[1]

The importance of DGER relates to findings in both animal and human studies that factors other than acid, namely bile and pancreatic enzymes, may play a significant role in mucosal injury and symptoms in patients with gastroesophageal reflux disease (GERD). However, the relative importance of acid and DGER to the development of esophageal mucosal injury is controversial and the subject of many studies using both animal models and human subjects.[2] Recent clinical studies, using state-of-the-art methods for detecting esophageal reflux of acid and duodenal contents are helping unravel the role of these potentially injurious agents in producing esophagitis. In this chapter, we will first briefly discuss the current data on the importance of acid and pepsin, and then extensively review the evidence for the importance of DGER in causing esophageal mucosal damage and symptoms.

## IMPORTANCE OF ACID AND PEPSIN

Substantial experimental and clinical evidence strongly supports the importance of acid and pepsin in causing esophageal mucosal injury. Animal studies[3,4] show that the esophageal mucosa is relatively resistant to reflux of acid alone unless it occurs at very high concentrations (pH 1.0–1.3). On the other hand, the combination of acid and even small concentrations of pepsin results in macroscopic, as well as microscopic, esophageal mucosal injury[4] (Figure 17.1).

The modern concept of "peptic esophagitis" was first suggested by Winkelstein in 1935[5] who proposed a role for gastric juice in the formation of esophagitis based upon clinical findings in five patients. However, studies by Aylwin et al.[6] were the first scientific evidence identifying the importance of acid and pepsin in the development of heartburn and esophageal mucosal injury. Using continuous esophageal aspiration in patients with hiatal hernia and esophagitis, they found that patients with esophagitis had aspirates of lower pH and higher pepsin concentration than those without esophagitis. Later, Tuttle et al.[7] measured the pH of the distal esophagus, finding that reflux of pH < 4 material coincided with the onset of heartburn, whereas a rise to a more neutral pH coincided with relief of symptoms.

Subsequently, a series of studies showed that patients with various grades of esophagitis, including Barrett's esophagus, have increased frequency and duration of esophageal exposure to pH < 4 refluxate[8–12]. Iascone et al[13] reported a direct relationship between the severity of esophageal mucosal injury and the degree and frequency of mucosal exposure to acid reflux. Later, studies by DeMeester et al.[14] found that over 90 percent of patients with esophagitis had increased amounts of acid reflux by 24-hour pH monitoring. The same group[15] reported that patients with Barrett's esophagus had significantly higher exposure times to pH < 4 than patients with esophagitis, who had higher exposure times than healthy controls, suggesting a significant role for acid reflux in the development of esophagitis and Barrett's esophagus.

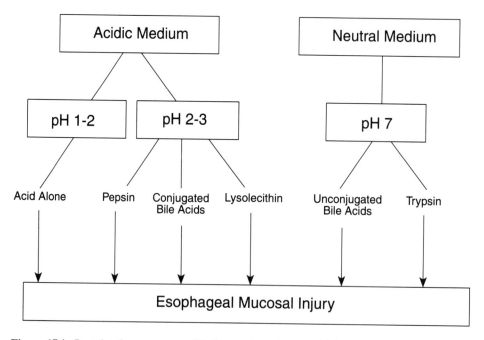

**Figure 17.1.** Postulated agents responsible for esophageal mucosal injury.

Separating the role of pepsin from acid in the production of esophagitis is difficult, since the optimum pH for the enzymatic activity of pepsin is below 3.[16] Studies show a positive correlation between the degree of abnormal acid and pepsin exposure and the severity of esophagitis. Bremner et al.[17] observed that patients with increased esophageal exposure to pH 1 to 2, corresponding to the known pKa of pepsin, had the most significant degrees of esophagitis. In a recent study, Gotley et al.[18] found that esophageal aspirates from patients with esophagitis have significantly higher concentrations of acid and pepsin than the aspirates from healthy controls. Furthermore, patients with Zollinger-Ellison syndrome, where the basal acid output is high and gastric pH favors optimum acidity for pepsin activity, have a 40–60 percent incidence of esophagitis despite normal or increased LES pressures.[19]

It is important to note that the frequency and duration of esophageal acid exposure is not always predictive of the degree of esophageal mucosal injury. This suggests the importance of other factors, including DGER, the inherent resistance of esophageal mucosa to acid injury and the role of saliva and bicarbonate-producing submucosal glands in the distal esophagus to neutralize refluxed acid.[20–23]

# IMPORTANCE OF DGER

Isolated case reports over the last 40 years suggest that duodenal contents alone, without acid or pepsin, may cause esophageal mucosal damage.[24–26] This was postulated by Helsingen, who observed that patients after gastrectomy with resulting achlorhydria could still develop severe esophagitis.[24] Others[25,26] also questioned the role of acid and pepsin as the sole causes of esophageal mucosal damage by reporting heartburn and esophagitis in patients with achlorhydria, with and without pernicious anemia.

*Animal Studies*

Esophageal mucosal damage by bile salts is dependent on the conjugation state of the bile acids and the pH of the refluxate. Using net acid flux across the rabbit esophageal lumen as an index of mucosal injury, Harmon et al.[27] showed that taurine-conjugated bile salts, taurodeoxycholate and taurocholate (both with pKa of 1.9) increased net acid flux at pH 2, whereas the unconjugated forms increased net acid flux at pH 7, but not at pH 2. Hence, conjugated bile acids are more injurious to the esophageal mucosa at acidic pH, although unconjugated bile acids are more harmful at pH 5–8. Similarly, Salo et al.[28] found that lysolecithin, a normal constituent of duodenal juice formed by pancreatic phospholipase A hydrolysis of lecithin in bile, causes histologic damage and alteration of the rabbit esophageal transmucosal potential difference in the presence of HCl, but there was no effect in the absence of HCl. Additionally, Kivilaakso et al.[29] showed that trypsin causes esophageal mucosal injury at pH 7.0, concluding that "alkaline" reflux esophagitis was caused by both unconjugated bile acids and trypsin at neutral pH values.

Therefore, as illustrated in Figure 17.1, there is evidence in the animal model for synergism between HCl and pepsin, as well as HCl and conjugated bile acids and lysolecithin, in causing esophageal mucosal damage. Similarly, unconjugated bile acids and trypsin are most injurious at pH 7.

*Human Studies*

Some groups have interpreted the above findings to suggest that aggressive acid suppression, although protective against the injurious effects of acid and possibly conjugated bile acid, may in fact perpetuate DGER, potentially causing complications in patients with GERD, such as Barrett's esophagus, dysplasia and adenocarcinoma.[9] However, the clinical importance of DGER in the absence of acid reflux has only recently been investigated. This may be because there is no "gold standard" for detecting DGER in humans.

## Methods for Measuring DGER

Various direct and indirect methods are employed for measuring DGER including endoscopy, aspiration studies (both gastric and esophageal), scintigraphy, ambulatory pH monitoring, and most recently, ambulatory bilirubin monitoring (Bilitec 2000). As summarized in Table 17.1, these tests have their strengths and shortcomings; however, reviewing some of the human studies using these tests can help us better appreciate the role of DGER in causing esophageal mucosal injury.

TABLE 17.1. ADVANTAGES AND DISADVANTAGES OF THE CURRENTLY AVAILABLE METHODS FOR DETECTING DUODENOGASTROESOPHAGEAL REFLUX (DGER)

| Method | Advantages | Disadvantages |
|---|---|---|
| Endoscopy | • Easy visualization of bile | • Poor sensitivity/specificity/ positive predictive value<br>• Requires sedation<br>• High cost |
| Aspiration studies | • Less invasive than endoscopy<br>• No sedation<br>• Low cost | • Short duration of study<br>• Requires familiarity with enzymatic assay for BA* |
| Scintigraphy | • Noninvasive | • Semiquantitative at best<br>• Radiation exposure<br>• High cost |
| pH monitoring | • Easy to perform<br>• Relatively noninvasive<br>• Prolonged monitoring<br>• Ambulatory | • pH > 7 not a marker for DGER<br>• Not specific for DGR |
| Bilirubin monitoring (Bilitec) | • Easy to perform<br>• Relatively noninvasive<br>• Prolonged monitoring<br>• Ambulatory<br>• Good correlation with gastric BA concentrations | • Current design underestimates DGER by about 30% in acidic medium (pH < 3.5)<br>• Requires modified diet |

*BA = bile acid

## Endoscopy

Bile is frequently seen in the stomach and esophagus of patients during endoscopy; however, studies indicate that this observation is a poor indicator of DGER.[30,31] Recently, Nasrallah et al.[30] evaluated 110 patients with bile stained gastric mucosa at endoscopy and found no correlation between the gastric bile acid concentrations, the degree of histologic injury, or the severity of endoscopic changes, suggesting that there was little clinical importance to bile stained mucosa at endoscopy. Similarly, using scintigraphy and gastric pH monitoring to assess DGER, Stein et al.[31] found poor sensitivity (37 percent), specificity (70 percent), and positive predictive value (55 percent) for endoscopy in the diagnosis of excessive DGER.

## Aspiration Techniques

One of the earliest methods used for evaluating DGER was the aspiration of gastric (or esophageal) contents with fluid analysis for bile acids. This technique allows direct detection of duodenal contents (bile acids and trypsin) with enzymatic or chromatographic measurements. Using this technique, recent studies[1,32] indicate that fasting bile acid concentrations may be increased in a graded fashion across the GERD spectrum, being highest among patients with Barrett's esophagus. However, the reports using aspiration techniques in detecting DGER may be criticized because of short aspiration periods and the limitations of the technique, in part because previous enzymatic measurements of bile acids, commonly studied in the postprandial periods, are now known to be inaccurate.[33]

## Scintigraphy

Scintigraphic studies show that DGER is a common phenomenon in normal individuals postprandially,[34] requiring that the evaluation of abnormal DGER be quantitative. Radionuclide techniques offer a noninvasive method for studying DGER; however, results are conflicting. Matikainen et al.[35] found no difference in the scintigraphic amount of DGER between 40 patients with esophagitis (10 percent scintigraphic reflux) and 150 healthy controls (14 percent scintigraphic reflux). On the other hand, Waring et al.[36] reported that patients with Barrett's esophagus, especially those with complicated Barrett's, had more frequent DGER detected by 99mTc DISIDA scintigraphy than healthy volunteers.

## Ambulatory Prolonged pH Monitoring

Until recently, the most popular method for detecting DGER was ambulatory 24-hour pH monitoring. Using this technique, Pellegrini et al.[37] introduced the term "alkaline" reflux, suggesting that pH > 7 be used as an indirect marker for DGER. Subsequently, Atwood et al.[38] reported that "alkaline" reflux was greater in patients with Barrett's esophagus when compared to patients with esophagitis or normal controls. Furthermore, they found that pH > 7 was significantly higher in complicated Barrett's patients (stricture, ulcer, dysplasia) than Barrett's patients without complications, whereas pH < 4 did not distinguish the two groups. Therefore, the authors suggested that prolonged exposure to duodenal contents alone may promote the development of complicated Barrett's esophagus and even adenocarcinoma.

However, the measurement of esophageal pH > 7 as a marker of DGER is confounded by several problems. Precautions must be taken to use only glass electrodes, a dietary restriction of foods with pH > 7, the inspection of patients for periodontal disease, and dila-

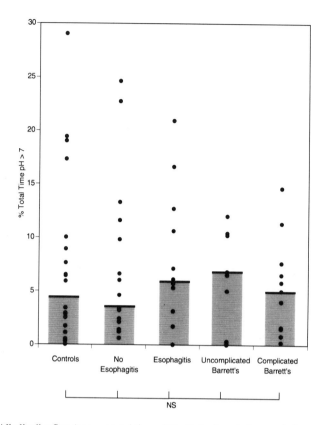

**Figure 17.2.** "Alkaline" reflux (percent total time pH > 7) for five study populations: controls, acid reflux patients with and without esophagitis, and patients with uncomplicated and complicated Barrett's esophagus. The quantity of "alkaline" reflux was similar across groups and did not distinguish them.

tion of strictures to avoid pooling of saliva. Additionally, Gotley et al.[39] found no relationship between "alkaline" exposure time and esophageal bile acids or trypsin. Similarly, Mattioli et al.,[23] using a triple-probe pH monitor placed in the distal esophagus, fundus and antrum, found that "alkaline" reflux, defined as a rise in pH > 7 from the antrum to the esophagus, was extremely uncommon. Singh et al.[21] and DeVault et al.[22] confirmed these observations by reporting that increased saliva production or bicarbonate production by the esophageal submucosal glands were the most common causes of esophageal pH > 7. Finally, using an ambulatory bilirubin monitoring device combined with pH monitoring, Vaezi et al.[40] reported no difference in the degree of percent total time pH > 7 between controls, patients with GERD, and those with Barrett's esophagus (Figure 17.2). Furthermore, Champion et al.[1] found no correlation between esophageal pH > 7 and bile reflux into the esophageal lumen (Figure 17.3), suggesting that the term "alkaline" reflux was a misnomer and should not be used when referring to DGER.

*Ambulatory Bilirubin Monitoring (Bilitec 2000)*

Recently, a new fiberoptic spectrophotometer (Bilitec 2000, Synectics, Stockholm, Sweden) was developed which detects DGER in an ambulatory setting, independent of pH

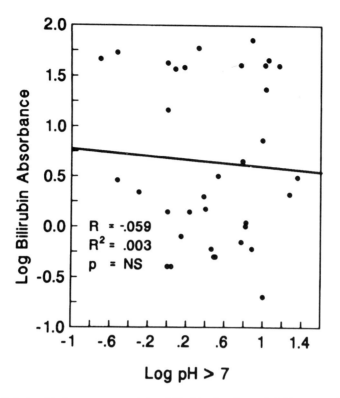

**Figure 17.3.** Relationship between percent time bilirubin absorbance ≥0.14 as a marker of bile reflux and esophageal pH > 7 in a group of healthy controls, patients with GERD and those with Barrett's esophagus.

(Figure 17.4).[41] This system utilizes the optical properties of bilirubin, the most common pigment in bile. Bilirubin has a characteristic spectrophotometric absorption band at 450 nm (Figure 17.5). The basic working principal of the system is that an absorption near this wavelength implies the presence of bilirubin and, therefore, represents DGER.

The system consists of a miniaturized fiberoptic probe which carries light signals into the probe tip and back to the optoelectronic system via a plastic fiberoptic bundle. The Teflon probe head is 9.5 mm in length and 4 mm in diameter. There is a 2.0 mm open groove in the probe across which two wavelengths of light are emitted and material sampled. Two light-emitting diodes at 470 and 565 nm represent the sources for the measurement of bilirubin and the reference signals, respectively (Figure 17.5). The portable photodiode system converts the light into an electrical signal. After amplification, the signals are processed by an integrated microcomputer, and the difference in absorption between the two diodes is calculated, representing bilirubin absorption in the samples of DGER. The period between two successive pulses from the same source, representing sampling time, is 8 seconds. In addition, the software averages between the absorbances calculated over two successive samplings in order to decrease the noise of the measurements. A total of 5,400 sample recordings may be stored during a 24-hour period.

DGER data are usually measured as percent time bilirubin absorbance ≥0.14 and can

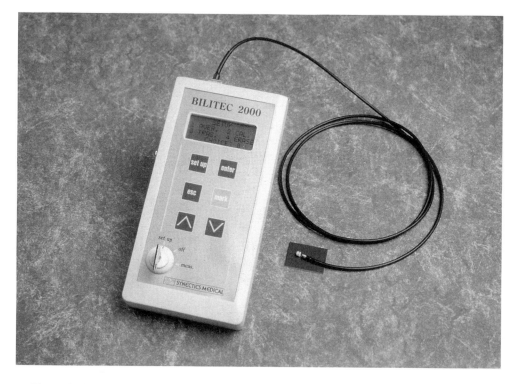

**Figure 17.4.** Bilitec 2000.

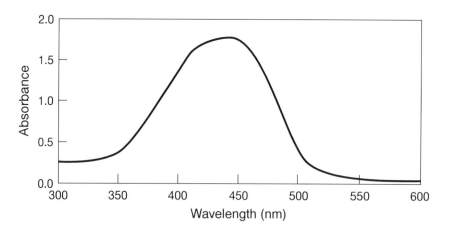

**Figure 17.5.** Spectrophotometric absorbance property of bilirubin.

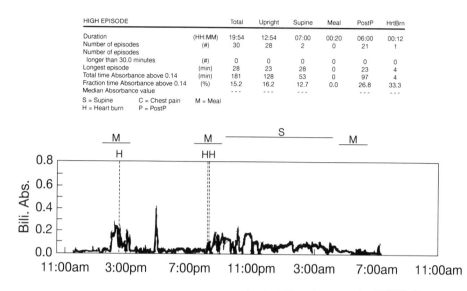

| HIGH EPISODE | | Total | Upright | Supine | Meal | PostP | HrtBrn |
|---|---|---|---|---|---|---|---|
| Duration | (HH:MM) | 19:54 | 12:54 | 07:00 | 00:20 | 06:00 | 00:12 |
| Number of episodes | (#) | 30 | 28 | 2 | 0 | 21 | 1 |
| Number of episodes | | | | | | | |
|   longer than 30.0 minutes | (#) | 0 | 0 | 0 | 0 | 0 | 0 |
| Longest episode | (min) | 28 | 23 | 28 | 0 | 23 | 4 |
| Total time Absorbance above 0.14 | (min) | 181 | 128 | 53 | 0 | 97 | 4 |
| Fraction time Absorbance above 0.14 | (%) | 15.2 | 16.2 | 12.7 | 0.0 | 26.8 | 33.3 |
| Median Absorbance value | | - - - | - - - | - - - | | - - - | - - - |

S = Supine    C = Chest pain    M = Meal
H = Heart burn    P = PostP

**Figure 17.6.** A typical tracing and data generated by the Bilitec in measuring DGER. Data are typically reported in percent time bilirubin absorbance $\geq 0.14$ (total, upright or supine).

be analyzed separately for total, upright, and supine periods (Figure 17.6). Percent bilirubin absorbance $\geq 0.14$ is commonly chosen as a cutoff because studies show that values below this number represent scatter due to suspended particles and mucus present in the gastric contents.[42] In a recent study[40] using 20 healthy controls, the 95th percentile values for percent total, upright and supine times bilirubin $\geq 0.14$ were 1.8 percent, 2.2 percent, and 1.6 percent respectively (Figure 17.7).

Studies from Dr. Paulo Bechi's laboratory[41] as well as our laboratory,[42] show a good correlation between Bilitec readings and bile acid concentration measurements of gastric aspirates using enzymatic assays (R = 0.71, $p < 0.01$ and R = 0.82, $p < 0.001$, respectively) (Figure 17.8). Furthermore, our studies show that Bilitec readings correspond to bile acid concentrations in the range of 0.01–0.60 mM, which are more representative of bile acid concentrations found in the human stomach (0.1–1.0 mM).

Due to limitations inherent to current Bilitec model, it is only a semiquantitative means of detecting DGER. Validation studies by Vaezi et al.[42] found that this instrument underestimates bile reflux by least 30 percent in an acidic medium (pH < 3.5). In solutions with pH < 3.5, bilirubin undergoes monomer to dimer isomerization which is reflected by the shift in the absorption wavelength from 453 nm to 400 nm (Figure 17.9). Since, Bilitec readings are based on the detection of absorption at 470 nm, this shift results in underestimation of the degree of DGER. Therefore, Bilitec measurements of DGER must always be accompanied by the simultaneous measurement of esophageal acid exposure using prolonged pH monitoring. Furthermore, a variety of substances may result in false positive readings by the Bilitec, since it indiscriminately records any substance absorbing around 470 nm. This necessitates use of a modified diet to avoid interference and false readings.[41] Finally, it is important to remember that Bilitec measures reflux of bilirubin and not bile acids; thereby presuming that the presence of bilirubin in the refluxate is accompanied by other duodenal contents. Although this is true in most cases, a few uncommon medical conditions (Gilbert's and Dubin-Johnson syndromes) may result in disproportionate secretion of bilirubin as compared to other duodenal contents, especially bile acids.

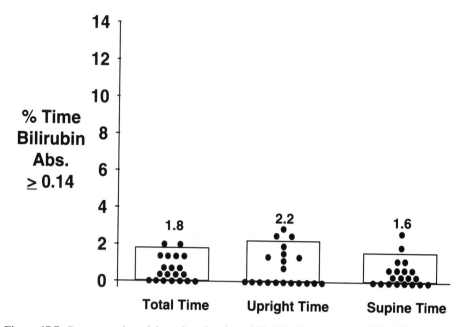

**Figure 17.7.** Percent total, upright and supine times bilirubin absorbance ≥0.14 in 20 normal subjects. Values within the boxed-in areas represent the 95th percentile of the normal range.

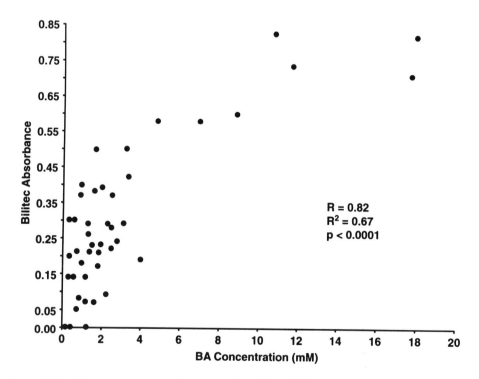

**Figure 17.8.** Bilitec absorbance readings and gastric bile acid (BA) concentrations.

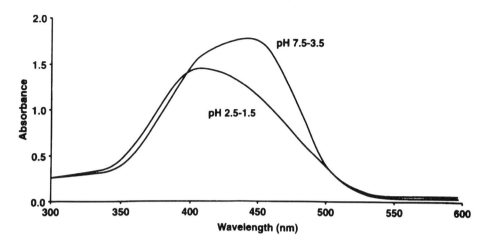

**Figure 17.9.** Spectrophotometric absorbance of bilirubin at different pH values.

# CLINICAL STUDIES USING THE BILITEC

### Case Study No. 1

A forty-nine-year-old businessman presented to our gastroenterology clinic with a two-year history of heartburn, regurgitation and excessive belching. Endoscopy showed a 3 cm hiatal hernia and a 7 cm segment of Barrett's esophagus. Esophageal manometry revealed a lower esophageal sphincter pressure of 6 mm Hg (normal value being higher than 10) Simultaneous 24-hour esophageal bilirubin and pH monitoring showed abnormal % total time pH < 4 (18 percent and percent total time bilirubin absorbance ≥0.14 (8 percent) (Figure 17.10).

Despite its limitations, Bilitec is an important advancement in the assessment of DGER in the clinical arena. Several studies using this new device are providing important insights into the role of DGER in causing esophageal mucosal injury in humans. Recently, Champion et al.[1] found a significant graded increase in *both* acid and DGER from controls to esophagitis patients, with the highest values observed in patients with Barrett's esophagus (Figure 17.11). Similar observation were made by Vaezi et al.[32] in patients with and without complications of Barrett's esophagus. They found that both groups of Barrett's patients had significantly greater quantities of acid and DGER than controls. More importantly, reflux of acid paralleled DGER and both were significantly higher in patients with complicated Barrett's than the uncomplicated group. The results of these two studies were recently confirmed by two other independent groups.[43,44] Furthermore, studies by Vaezi et al.[40] found that simultaneous esophageal exposure to both acid and DGER was the most prevalent reflux pattern occurring in 95 percent of patients with Barrett's esophagus and 79 percent of GERD patients (Figure 17.12). Thus, these studies support the findings in animals, suggesting a possible synergy between acid and DGER in the development of esophagitis and Barrett's esophagus.

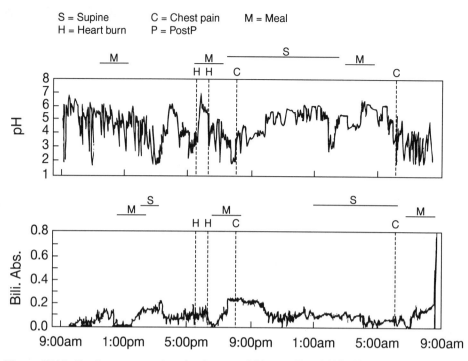

**Figure 17.10.** Tracings representing simultaneous 24-hour pH and bilirubin monitoring in a patient with Barrett's esophagus.

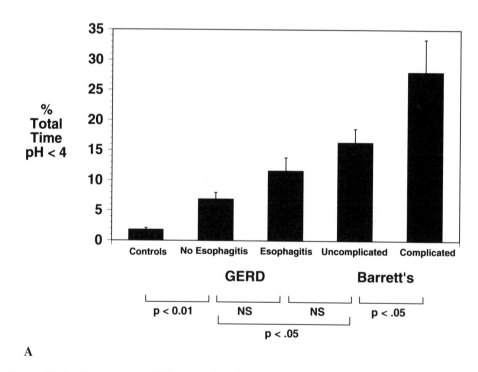

A

**Figure 17.11.** Group mean ± SE for (A) acid reflux and (B) DGER for the four study populations.

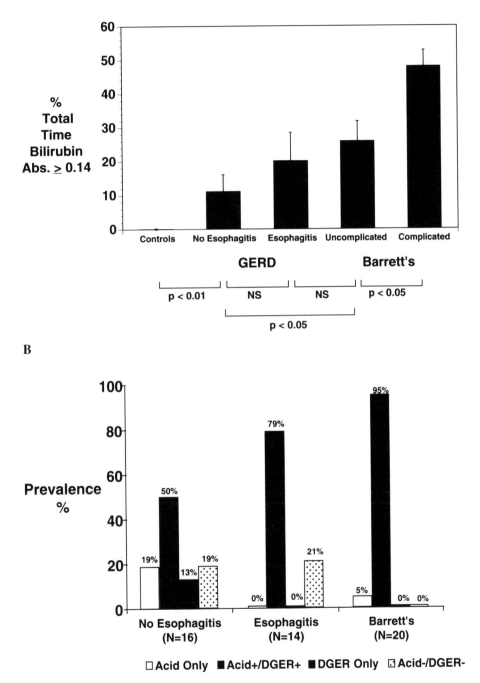

**Figure 17.12.** Prevalence of esophageal exposure to acid and DGER in the GERD subgroups. Esophageal exposure to both acid and DGER occurred in 50 percent of patients without esophagitis, 79 percent of patients with esophagitis, 95 percent of patients with Barrett's esophagus.

**Case Study No. 2**

A sixty-five-year-old Vietnam veteran presents with a 15-year history of intermittent nausea, vomiting and regurgitation of bitter bilious material. His symptoms started about two years after he underwent a partial gastrectomy with Bilroth II anstomosis for peptic ulcer disease. Endoscopy showed a normal appearing esophagus without esophagitis. The stomach was bathed with a large amount of bile stained fluid and the gastric mucosa was reddened and friable. The LES pressure was low at 10 mm Hg. Simultaneous 24-hour esophageal pH and bilirubin monitoring showed normal esophageal exposure to acid (percent total time pH < 4 of 0.1 percent but abnormally high amounts of DGER (percent total time bilirubin absorbance ≥0.14 of 27.9 percent) (Figure 17.13).

The role of DGER in producing esophageal mucosal injury, in the absence of acid reflux, was not clarified until recently. Sears et al.[45] studied 13 partial gastrectomy patients with reflux symptoms finding increased DGER by Bilitec monitoring in 77 percent of patients. However, endoscopic esophagitis was only present in those with concomitant acid reflux. Additionally, Vaezi et al.[46] observed that 24 percent of upper GI symptoms reported by partial-gastrectomy patients was due to DGER in the absence of acid reflux. Therefore, these studies underscore the important point that DGER, without excessive acid reflux, can cause reflux symptoms but does not usually produce esophagitis.

**Treatment-Case Studies Nos. 1 and 2**

The forty-nine-year-old business man with Barrett's esophagus was treated with omeprazole 20 mg once daily. In the one-month clinic follow-up, his symptoms of heartburn and regurgitation had resolved. The sixty-five-year-old Vietnam veteran was treated with cisapride 20 mg qid with resolution of his symptoms.

Bilitec is also used to study the effects of drug therapy on DGER. Studies by Champion et al.[1] found that aggressive acid suppression with omeprazole (20 mg BID) dramatically decreased *both* acid and DGER in patients with severe GERD (Figure 17.14). Although not specifically studied, the authors speculated this was due to omeprazole's inhibition of both gastric acidity and volume. This finding has important implications for treating patients with both acid and DGER, suggesting that medical therapy may decrease both constituents to a similar degree to antireflux surgery. Furthermore, the higher intragastric and intraesophageal pH environment created by the proton pump inhibitors inactivate conjugated bile acids, the main DGER ingredients implicated in causing esophageal mucosal injury.[47] In partial-gastrectomy patients having upper GI symptoms due to non-acidic DGER, a recent randomized double-blind cross-over study[48] found that cisapride (20mg qid) significantly reduces both DGER measured by the Bilitec and the associated upper GI. Thus, medical therapy with this promotility drug is an alternative to surgical Roux-en-Y diversion in this difficult group of patients.

# SUMMARY

Esophageal pH monitoring and pH > 7 is a poor marker for reflux of duodenal contents into the esophagus. Therefore, the term "alkaline reflux" should no longer be used in re-

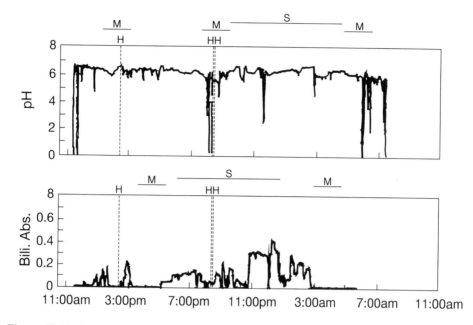

**Figure 17.13.** Tracings representing simultaneous 24-hour pH and bilirubin monitoring in a post-gastrectomy patient with upper GI symptoms.

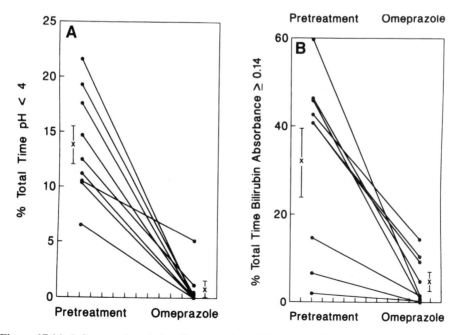

**Figure 17.14.** Influence of marked acid suppression with omeprazole (20 mg twice daily) on acid and DGER in nine patients with GERD.

ferring to this process. Bilitec is the method of choice in detecting DGER and should always be used simultaneously with esophageal pH monitoring for acid reflux. DGER in nonacidic environments (i.e. partial-gastrectomy patients) may cause symptoms but does not cause esophageal mucosal injury. Acid and duodenal contents usually reflux into the esophagus simultaneously, and may be contributing to the development of Barrett's metaplasia and possibly adenocarcinoma. Proton pump inhibitors decrease both acid and DGER by reducing the volume of gastric contents available to reflux into the esophagus and raise intragastric pH. The promotility agent cisapride decreases DGER by increasing LES pressure and improving gastric emptying.

# REFERENCES

1. Champion G, Richter JE, Vaezi MF, Singh S, Alexander R. Duodenogastroesophageal reflux: Relationship to pH and importance in Barrett's esophagus. *Gastroenterology* 107:747–754, 1994.
2. Vaezi MF, Singh S, Richter JE. Role of acid and duodenogastric reflux in esophageal mucosal injury: A review of animal and human studies. *Gastroenterology* 108:1897–1907, 1995.
3. Redo SF, Barnes WA, de la Sierra AO. Perfusion of the canine esophagus with secretions of the upper gastro-intestinal tract. *Ann. Surg.* 149:556–564, 1959.
4. Goldberg HI, Dodds WJ, Gee S, Montgomery C, Zboralske FF. Role of acid and pepsin in acute experimental esophagitis, *Gastroenterology* 56:223–230, 1969.
5. Winkelstein A. Peptic esophagitis. A new clinical entity. *JAMA* 104:906–909, 1935.
6. Aylwin JA. The physiological basis of reflux esophagitis in sliding distal diaphragmatic hernia. *Thorax* 8:38–45, 1953.
7. Tuttle SG, Bettarello A, Gossman MI. Esophageal acid perfusion test and a gastroesophageal reflux in patients with esophagitis. *Gastroenterology* 38:861–72, 1960.
8. Zamost BJ, Hirschberg J, Ippoliti AF. Esophagitis in scleroderma: Prevalence and risk factors. *Gastroenterology* 92:421–428, 1987.
9. Stein HJ, Barlow AP, DeMeester TR, Hinder RA. Complications of gastroesophageal reflux disease. *Ann. Surg.* 216:35–43, 1992.
10. Hennessy TPJ. Barrett's esophagus. *Br. J. Surg.* 72:336–340, 1985.
11. Gillen P, Keeling P, Byrne PJ, Hennessy TPJ. Barrett's esophagus: pH profile. *Br. J. Surg.* 74:774–776, 1987.
12. Stein HJ, Siewert JR. Barrett's esophagus: Pathogenesis, epidemiology, functional abnormalities, malignant degeneration, and surgical management. *Dysphagia* 8:276–288, 1993.
13. Iascone C, DeMeester TR, Little AG, Skinner DB. Barrett's esophagus: Functional assessment, proposed pathogenesis, and surgical therapy. *Arch. Surg.* 118:543, 1983.
14. DeMeester TR, Wernly JA, Little AG, Bermudez G, Skinner DB. Technique, indications, and clinical use of 24 hour esophageal pH monitoring. *J. Thorac. Cardiovasc. Surg.* 79:656–670, 1980.
15. Stein HJ, Hoeft S, DeMeester TR. Reflux and motility pattern in Barrett's esophagus. *Dis. Esoph.* 5:21–28, 1992.
16. Lillemoe KD, Johnson LE, Harmon JW: Alkaline esophagitis: A comparison of the ability of components of gastroduodenal contents to injure rabbit mucosa. *Gastroenterology* 85:621–628, 1983.
17. Bremner RM, Crookes PF, DeMeester TR, Peters JH, Stein H: Concentration of refluxed acid and esophageal mucosal injury. *Am J Surg* 164:522–527, 1992.
18. Gotley DC, Morgan AP, Ball D, Owens RW, Cooper MJ: Composition of gastro-oesophageal refluxate. *Gut* 32:1093–1099, 1991.

19. Miller LS, Fruacht H, Saeed ZA, et al: Esophageal involvement in the Zollinger Ellison syndrome (ZES). *Gastroenterology* 98:341–345, 1990.

20. Hamilton BH, Orlado RC. In vivo alkaline secretion by mammalian esophagus. *Gastroenterology* 97:640–648, 1989.

21. Singh S, Bradley LA, Richter JE: Determinants of oesophageal "alkaline" pH environment in controls and patients with gastro-oesophageal reflux disease. *Gut* 34:309–316, 1993.

22. Devault KR, Georgeson S, Castell DO: Salivary stimulation mimics esophageal exposure to refluxed duodenal contents. *Am J Gastroenerol* 88:1040–1043, 1993.

23. Mattioli S, Pilotti V, Felice V, Lazzzari A, Zannoli R, Bachi ML, et al: Ambulatory 24 hour pH monitoring of the esophagus, fundus and antrum. *Dig Dis Sci* 35:929–938, 1990.

24. Helsingen N: Oesophagitis following total gastrectomy. A follow up study on 9 patients 5 years or more after operation. *Acta Chir Scand* 118:190–201, 1959.

25. Orlando RC, Bozymski EM: Heartburn in pernicious anemia-A consequence of bile reflux. *N Engl J Med* 289:522–523, 1973.

26. Palmer ED: Subacute erosive ("peptic") esophagitis associated with achlorhydria. *N Engl J Med* 262:927–929, 1960.

27. Harmon JW, Johnson LF, Maydonovitch CL: Effects of acid and bile salts on the rabbit esophageal mucosa. *Dig Dis Sci* 26:65–72, 1981.

28. Salo J, Kivilaakso E: Role of luminal $H^+$ in the pathogenesis of experimental esophagitis. *Surgery* 92:61–68, 1982.

29. Kivilaakso E, Fromm D, Silen W. Effect of bile salts and related compounds on isolated esophageal mucosa. *Surgery* 87:280–285, 1980.

30. Nasrallah SM, Johnston GS, Gadacz TR, Kim KM. The significance of gastric bile reflux seen at endoscopy. J Clin Gastroenterol 9:514–517, 1987.

31. Stein HJ, Smyrk TC, DeMeester TR, Rouse J, Hinder RA: Clinical value of endoscopy and histology in the diagnosis of duodenogastric reflux disease. *Surgery* 112:796–804, 1992.

32. Vaezi MF, Richter JE: Synergism of acid and duodenogastroesophageal reflux in complicated Barrett's esophagus. *Surgery* 117:699–704, 1995.

33. Mittal RK, Reuben A, Whitney JO, McCallum RW: Do bile acids reflux into the esophagus? A study in normal subject and patients with GERD. *Gastroenterology* 92:371–375, 1987.

34. Muller-Lissner SA, Fimmel CJ, Sonnenberg A: Novel approach to quantify duodenogastric reflux in healthy volunteers and in patients with type I gastric ulcer. *Gut* 24:510–518, 1983.

35. Matikainen M, Taavitsainen M, Kalima TV: Duodenogastric reflux in patients with heartburn and esophagitis. *Scand J Gastroenterol* 16:253–55, 1981.

36. Warring JP, Legrand J, Chinichian A, Sanowski RA: Duodenogastric reflux in patients with Barrett's esophagus. *Dig Dis Sci* 35:759–762, 1990.

37. Pellegrini CA, DeMeester TR, Wernly JA, Johnson LF, Skinner DB: Alkaline gastroesophageal reflux. *Am J Surg* 75:177–184, 1978.

38. Attwood SEA, Ball CS, Barlow AP, Jenkinson L, Norris TL, Watson A: Role of intragastric and intraoesophageal alkalinization in the genesis of complications in Barrett's columnar lined lower oesophagus. *Gut* 34:11–15, 1993.

39. Gotley DC, Appleton GVN, Cooper MJ: Bile acids and trypsin are unimportant in alkaline esophageal reflux. *J. Clin. Gastroenterol.* 14:2–7, 1992.

40. Vaezi MF, Richter JE. Role of acid and duodenogastroesophageal reflux in gastroesophageal reflux disease. *Gastroenterology* 111:1192–99, 1996.

41. Bechi P, Paucciani F, Baldini F, Cosi F, Falciai R, Mazzanti R. Castagnli A, Pesseri A, Boscherini S. Long-term ambulatory enterogastric reflux monitoring. Validation of a new fiberoptic technique. *Dig Dis Sci* 38:1297–1306, 1993.

42. Vaezi MF, LaCamera RG, Richter JE: Bilitec 2000 ambulatory duodenogastric reflux monitoring system. Studies on its validation and limitations. *Am J Physiol* 267:G1050–G1057, 1994.

43. Kauer WK, Peters JH, DeMeester TR, et al: Mixed reflux of gastric and duodenal juice is more harmful to the esophagus than gastric juice alone. *Ann Surg* 222:525–533, 1995.

44. Caldwell MTP, Lawlor P, Byrne PJ, et al: Ambulatory oesophageal bile reflux monitoring in Barrett's oesophagus. *Br J Surg* 82:657–660, 1995.

45. Sears RJ, Champion G, Richter JE: Characteristics of partial gastrectomy (PG) patients with esophageal symptoms of duodenogastric reflux. *Am J Gastroenterol* 90:211–215, 1995.

46. Vaezi MF, Richter JE: Acid and duodenogastroesophageal reflux in postgastrectomy patients: Response to therapy. *Am J Gastroenterol* 90:A80. 1995.

47. Harmon JW, Johnson LF, Maydonovitch CL: Effects of acid and bile salts on the rabbit esophageal mucosa. Dig. Dis. Sci. 26:65–72, 1981.

48. Vaezi MF, Sears R, Richter JE: Double-blind placebo-controlled cross-over trial of cisapride in postgastrectomy patients with duodenogastric reflux. *Dig Dis Sci* 41:754–763, 1996.

# Index

Note: Page numbers in *italics* refer to illustrations; page numbers followed by t refer to tables.